AMERICAN POLITICAL, ECONOMIC, AND SECURITY ISSUES

# FACES OF ALASKA: LAND COVER, RENEWABLE ENERGY AND NATURAL GAS ISSUES

# American Political, Economic, and Security Issues

Additional books in this series can be found on Nova's website under the Series tab.

Additional E-books in this series can be found on Nova's website under the E-books tab.

# Energy Science, Engineering and Technology

Additional books in this series can be found on Nova's website under the Series tab.

Additional E-books in this series can be found on Nova's website under the E-books tab.

AMERICAN POLITICAL, ECONOMIC, AND SECURITY ISSUES

# FACES OF ALASKA: LAND COVER, RENEWABLE ENERGY AND NATURAL GAS ISSUES

MARK C. MONTEITH
EDITOR

Nova Science Publishers, Inc.
*New York*

For permission to use material from this book please contact us:
Telephone 631-231-7269; Fax 631-231-8175
Web Site: http://www.novapublishers.com



**LIBRARY OF CONGRESS CATALOGING-IN-PUBLICATION DATA**

Faces of Alaska : land cover, renewable energy, and natural gas issues /
editor, Mark C. Monteith.
p. cm.
Includes index.
ISBN 978-1-61122-733-8 (hardcover)
1. Land cover--Alaska. 2. Land use--Alaska. 3. Renewable energy
resources--Alaska. 4. Natural gas--Alaska. I. Monteith, Mark C.
GF504.A4F33 2010
333.709798--dc22
2010041360

*Published by Nova Science Publishers, Inc. + New York*

# CONTENTS

# PREFACE

This book explores the land cover, renewable energy and natural gas issues facing Alaska today. The land surface of Alaska is sparsely populated and the impacts from humans are far less extensive when compared to the contiguous United States. A brief survey of renewable energy technologies applicable to Alaska's climate, latitude, geography, and geology is also discussed as are Alaska's natural renewable energy resources and which renewable energy technologies would be most productive. This book also provides a review of the history of efforts to develop an Alaskan natural gas pipeline, including project status, recent developments, and the current project outlook.

Chapter 1 - The purpose of this report is to provide statewide baseline information on the status and potential drivers of land-cover change in Alaska. The information gathered for this report is based on a review and analysis of published literature and consists of prominent factors contributing to the current state of the land surface of Alaska as well as a synthesis of information about the status and trends of the factors affecting the land surface of Alaska. The land surface of Alaska is sparsely populated and the impacts from humans are far less extensive when compared to the contiguous United States. The changes in the population and the economy of Alaska have historically been driven by boom and bust cycles, primarily from mineral discoveries, logging, military expansion, and oil and gas development; however, the changes as a result of these factors have occurred in relatively small, localized areas. Many of the large-scale statewide changes taking place in the land surface however, are a result of natural or climate driven processes as opposed to direct anthropogenic activities. In recent times, reports such as this have become increasingly useful as a means of synthesizing information about the magnitude and frequency of changes imparted by natural and anthropogenic forces. Thus, it is essential to assess the current state of the land surface of Alaska and identify apparent trends in the surficial changes that are occurring in order to be prepared for the future.

Chapter 2 - This paper delivers a brief survey of renewable energy technologies applicable to Alaska's climate, latitude, geography, and geology. We first identify Alaska's natural renewable energy resources and which renewable energy technologies would be most productive. We survey the current state of renewable energy technologies and research efforts within the U.S. and, where appropriate, internationally. We also present information on the current state of Alaska's renewable energy assets, incentives, and commercial enterprises. Finally, we describe places where research efforts at Sandia National Laboratories could assist the state of Alaska with its renewable energy technology investment efforts

Chapter 3 - Constructing a natural gas pipeline from Alaska's North Slope to the lower-48 states has been a government priority—periodically—for more than four decades. Beginning with the Alaska Natural Gas Transportation Act of 1976, Congress has repeatedly affirmed a national need for an Alaska gas pipeline. In remarks to the press President Obama has likewise described an Alaska gas pipeline as "a project of great potential…as part of a comprehensive energy strategy."

In: Faces of Alaska
Editor: Mark C. Monteith

ISBN: 978-1-61122-733-8


*Chapter 1*

# THE FACE OF ALASKA: A LOOK AT LAND COVER AND THE POTENTIAL DRIVERS OF CHANGE

***Benjamin M. Jones***

## ABSTRACT

The purpose of this report is to provide statewide baseline information on the status and potential drivers of land-cover change in Alaska. The information gathered for this report is based on a review and analysis of published literature and consists of prominent factors contributing to the current state of the land surface of Alaska as well as a synthesis of information about the status and trends of the factors affecting the land surface of Alaska. The land surface of Alaska is sparsely populated and the impacts from humans are far less extensive when compared to the contiguous United States. The changes in the population and the economy of Alaska have historically been driven by boom and bust cycles, primarily from mineral discoveries, logging, military expansion, and oil and gas development; however, the changes as a result of these factors have occurred in relatively small, localized areas. Many of the large-scale statewide changes taking place in the land surface however, are a result of natural or climate driven processes as opposed to direct anthropogenic activities. In recent times, reports such as this have become increasingly useful as a means of synthesizing information about the magnitude and frequency of changes imparted by natural and anthropogenic forces. Thus, it is essential to assess the current state of the land surface of Alaska and identify apparent trends in the surficial changes that are occurring in order to be prepared for the future.

## INTRODUCTION

Alaska is the largest state in the Nation. It spans a land area of around 1,500,000 km$^2$. This corresponds to almost one-fifth the size of the conterminous United States (Figure 1). The northernmost point in Alaska is Point Barrow (71°23'N, 156°29'W) and the southernmost point is the southern tip of Amatignak Island (51°15'N, 179°06'W), located in the Aleutian Island arc. When considering the east to west expanse of Alaska, the eastern extreme is near Portland Canal in Southeast Alaska (55°00'N, 130°00"W) and the western extreme is Cape Wrangell, Attu Island (52°55'N, 172°27'E). Thus, Alaska spans an area roughly 20° in latitude by 58° in longitude. Alaska crosses the International Date Line, thus with respect to a global geographic perspective that is referenced from the prime meridian in Greenwich, England, it contains the western most location in the United States, Amatignak Island (51°1 5'N, 179°06'W), as well as the eastern most location in the United States, Pochnoi Point, Semisopochnoi Island (51°57'N, 179°47'E) (Figure 2).

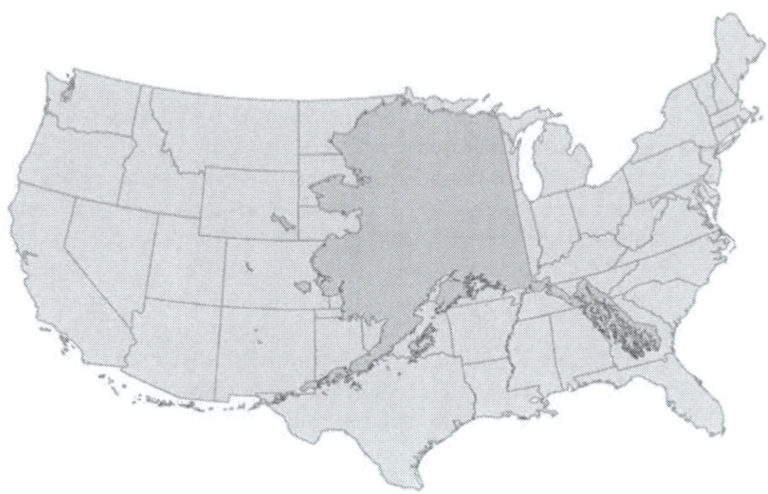

Figure 1. Comparison of Alaska to the contiguous United States. Alaska is nearly one-fifth the size of the lower 48 states.

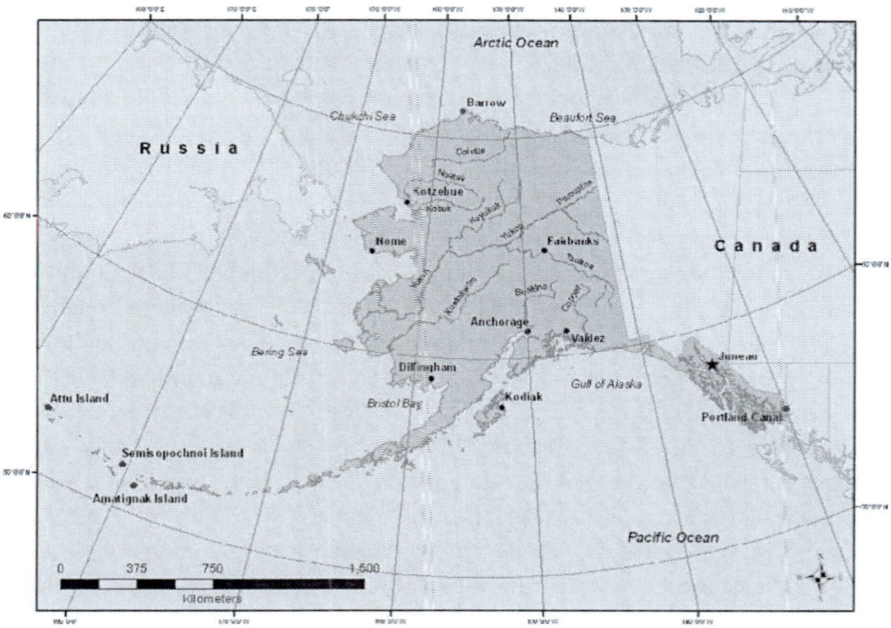


Figure 2. General geography of Alaska showing large rivers, major cities (black) and places mentioned in the text (red) (data source: Alaska Geospatial Data Clearinghouse).

Alaska is bounded to the east by Canada (a border of 2,475 km), to the north by the Arctic Ocean (Beaufort and Chukchi Sea), to the west by the Bering Sea, and to the south by the Pacific Ocean (Figure 2). Measured in straight distance, the coastline of Alaska is 10,686 km in length. However, if each of the bays, fjords, islands and channels were considered, the length of the coastline of Alaska would be roughly 54,500 km, almost three times that of the conterminous United States.

The average elevation of Alaska is around 580 meters above sea level, but the topography across the state is extremely variable (Figure 3). It ranges from sea level along the Arctic Ocean, Bering Sea, and Pacific Ocean to the summit of Denali (Mt. McKinley) at an elevation of 6,194 meters above sea level. Denali (Mt. McKinley) represents the highest point in North America and it has the highest base-to-summit gain in elevation of any mountain on the surface of the earth, roughly 5,500 m. There are 39 mountain ranges found within the state and 17 of the 20 tallest peaks of the United States are found within these mountain ranges.

The climate across the state is also extremely variable; primarily controlled by elevation, latitude, and distance from the coast (Benson et. al., 1983). The northern region of the state experiences the coldest annual temperatures, however, the interior of the state experiences the greatest differences in temperature annually, and the southeastern region of the state receives the most precipitation.

Another characteristic of the state is the cryosphere or that part of the earth's surface that is perennially frozen (van Everdingen, 2005), which is controlled to a large extent by climate. In Alaska, the cryosphere consists of ice caps, ice sheets, glaciers, sea ice, lake ice, river ice, snow, and perennially frozen ground (permafrost). Two of the primary components of the cryosphere, glaciers and permafrost, have a large and direct influence on the land surface of Alaska. At present, glaciers cover about 75,000 $km^2$ of the surface of Alaska and account for nearly 5% of the total land surface (Dyurgerov and Meier, 1997). Most of the glaciers are found in the mountain ranges in the southern part of the state (Chugach Range, Wrangell- St. Elias Range, Alaska Range), while a few are located in the northern part of the state in the Brooks Range. About 82% of the land surface of Alaska is underlain by permafrost which is divided into four generalized categories of permafrost: continuous, discontinuous, sporadic, and isolated. Each of these characteristics refers to the lateral continuity of the permafrost in a regional context (Pewe, 1975). A more detailed description of permafrost is presented in the subsequent chapters.

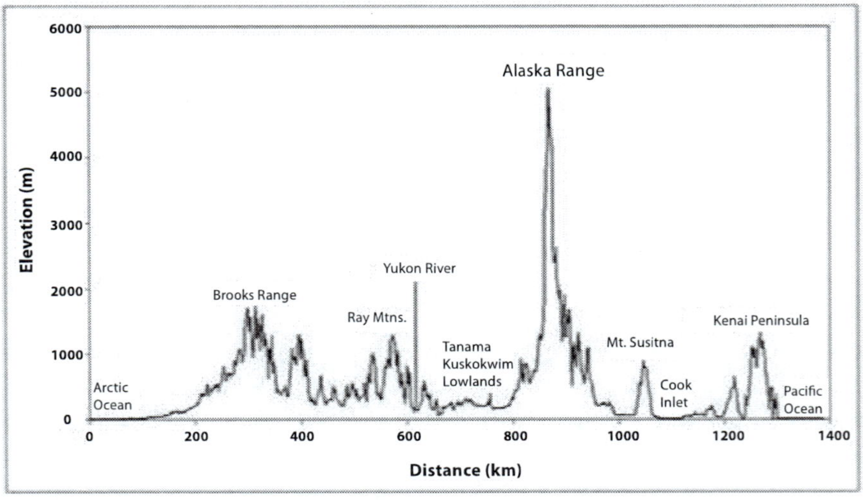


Figure 3. Elevation profile ranging from north to south through Alaska at ~151° W longitude.

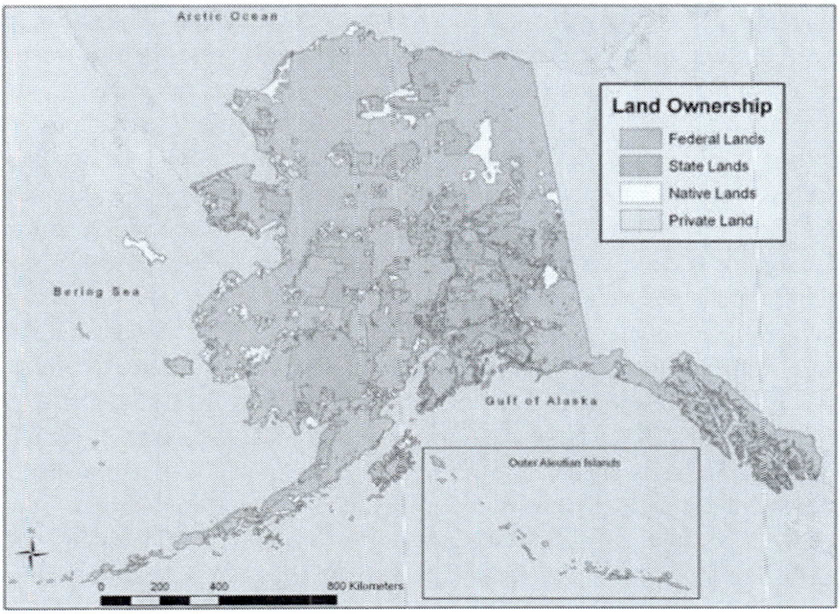


Figure 4. Alaska land status and ownership map. Federal lands constitute the majority of the state, followed by state lands, native lands, and privately owned lands (data source: Alaska Department of Natural Resources).

## LAND OWNERSHIP

Land ownership within the state has an influence on many of the potential changes that could occur on the land surface and a series of major events have contributed to the current ownership of land in Alaska. The first event occurred in 1867 when the United States government purchased the Alaska Territory from Russia for $70 million. The next major event occurred when Alaska became a state in 1959. At which time, the federal government granted 28% of the land to the state with the caveat that the land was not already designated as federal land, the state chose their sites based on its potential for settlement, potential for resource development, and its potential for recreational land (Alaska Department of Natural Resources, 2000).

In more recent times, two major congressional acts were passed that played a prominent role in the use of land in Alaska – ANSCA and ANILCA. In 1971 Congress passed the Alaska Native Claims Settlement Act (ANSCA).

This act granted village and native corporations the rights to 178,000 $km^2$ and nearly 1 billion dollars in start up funds. These funds were used to establish fixed settlements with modern day amenities; such as, sewage systems, power plants, and runways. The Alaska National Lands Act (ANILCA) was passed in 1980, establishing 220,000 $km^2$ of national wildlife refuge lands; placed portions of 25 rivers into the national wild and scenic river system; added 13,000 $km^2$ of national forest land; and 176,000 $km^2$ of new additions to the national park system. As a consequence, these areas were removed from the pool of land that the state could draw upon.

Overall, the ownership of land within Alaska can be divided into four groups: federal, state, native (private), and other private lands (Figure 4). Through the years following purchase, statehood, and ANILCA, the federal government remains the largest landowner in Alaska at 890,000 $km^2$ (or 60%). These federal lands include military reservations, national parks, national wildlife refuges, national forests, and the National Petroleum Reserve- Alaska (NPRA). The state owns the second largest proportion of land in Alaska 425,000 $km^2$ (or 28%), although, to date only about 85% has been allotted (Alaska Department of Natural Resources, 2000). The remainder of the land is considered to be in the private domain consisting of native corporation land (178,000 $km^2$ or 12%) and all other land owned by individuals (<1%).

# PHYSICAL ENVIRONMENT

## Ecoregions

The sheer vastness of Alaska combined with its latitudinal position, the differences in topography, and its juxtaposition between the frigid Arctic Ocean and the relatively warmer Pacific Ocean has created a landscape that is very diverse. Many of the surficial features change with distance from the coast, altitudinal position, and latitude. These are in a large part driven by variations in abiotic elements, such as the regional climate and the amount of solar insolation that the surface receives. Several different attempts have been made at the division of the state into distinct regions based on variations in climate and ecology (Nowacki et al., 2001; Status and Trends of the Nation's Biological Resources, 1998; Nowacki et al., 1995; Gallant et al., 1995). The most recent attempt by Nowacki et al. (2001) will be referred to in this report which identifies 32 created by using a "tri-archy" based on climate parameters, vegetation response, and disturbance processes (Nowacki et. al., 2001). The ecoregions are combined into three groups at the broadest scale (level 1) and eight groups at the moderate scale (level 2) and establishes a baseline for studying, managing, and understanding the ecosystems of Alaska and their driving processes (Figure 5).

The Level 1 ecoregions consist of Polar, Boreal and Maritime. The key features of the Polar level are non- forested terrain with a cold and dry climate and it covers 24.8% of the land area of Alaska. The Polar level includes the Arctic Tundra and Bering Tundra groups in the Level 2 classification. The Boreal level consists of the Intermontane Boreal, the Bering Taiga, and Alaska Range Transition and is the dominant Level 1 group in that it covers 56.9% of the surface. This level is identified by a forested landscape with a dry climate, a large seasonal variation in temperature and areas prone to forest fires. The third Level 1 group is the Maritime level. It is restricted to the southern portion of the state and it is the smallest of the Level 1 regions. It covers 18.3% of the land area and it includes the Level 2 groups of the Maritime (Coastal Rainforests, the Coastal Mountains Transition, and the Aluetian Meadows). The Maritime groups are characterized by forested landscapes with a warm and wet climate, limited variation in temperature throughout the season, and areas subject to high wind events.

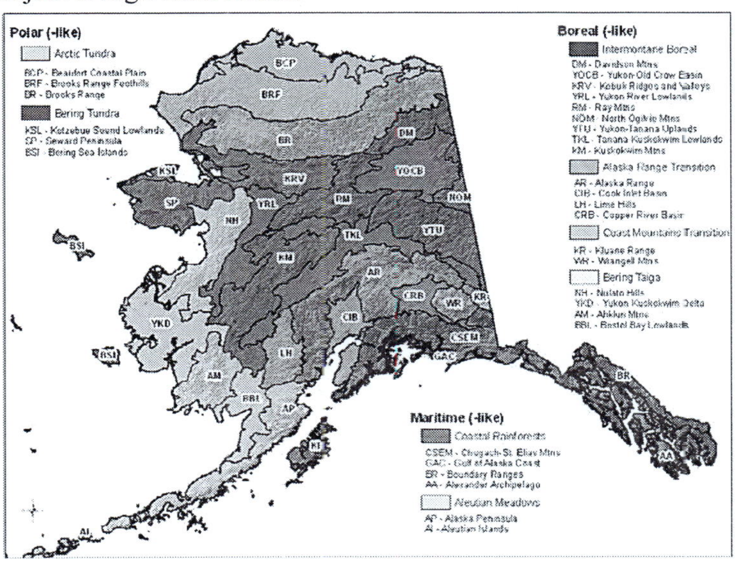


Figure 5. Unified Ecoregions of Alaska (modified from Nowacki et al. 2001; ecoregions have been clipped to Alaska border). Level 1 ecoregions consist of Polar, Boreal, and Maritime like. Level 2 ecoregions are distinguished by color and Level 3 ecoregions are abbreviated on the map and their meaning is conveyed in the legend.

## Forests

The distribution and extent of forests in Alaska are driven by geology, elevation, latitude, climate, insects, plant diseases and animal and human activities. The forests in Alaska are primarily restricted to the Boreal and Maritime ecoregions (Figure 6). Both of these areas however, exhibit a different type of forest community and are subject to different disturbance regimes.

The boreal forest of the interior of Alaska accounts for nearly 85% of the forested regions in the state. The tree species associated with the ecoregions of the interior are white spruce, black spruce, paper birch, quaking aspen, balsam poplar, and tamarack, with permafrost playing a major role in the presence and distribution of species across the region. Areas underlain by permafrost have cold soils that remain saturated throughout the summer with black spruce and tamarack typically occurring here. White spruce on the other hand is found on warmer soils with adequate drainage, such as slopes with a southern exposure and along rivers. The hardwood trees are primarily the pioneer species and are typically found in recently disturbed areas. The dominant disturbance regime in the boreal forest ecosystem is fire which is predominantly a result of lightening strikes.

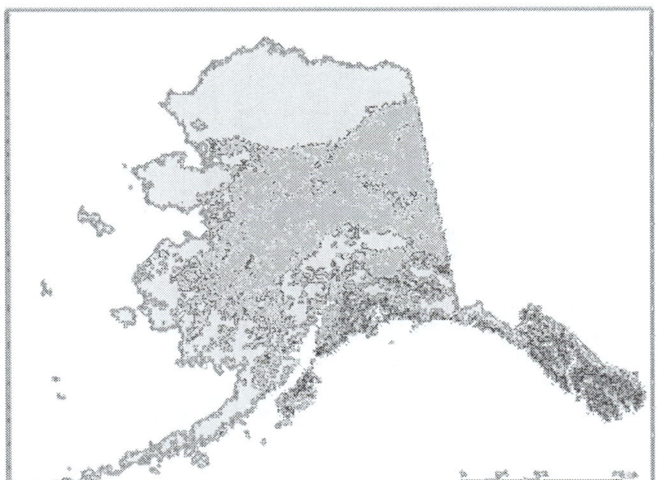

Figure 6. Forested area in Alaska (excluding shrubs). The Boreal Ecoregion contains nearly 85% of the forested area (light green) in the state and the Maritime Ecoregion contains around 13% of the forested area (dark green) (data source: Fleming, 1991; Forest Health Monitoring Clearinghouse).

The forests associated with the Maritime ecoregion consist of the Coastal Rainforest and the Coastal Mountain Transition. The coastal rainforest region is a continuation of the temperate rain forest that extends northward from Washington and SW Canada. The most predominant tree species in this region is the western hemlock; however, the overall forest composition of this region varies from south to north. The forests in the southern portion of this region are a combination of western hemlock, sitka spruce, pacific silver spruce, and pacific yew. Whereas to the north, western hemlock becomes dominant, pacific silver and pacific yew are no longer present, and there is a greater abundance of western red cedar and Alaska cedar. The forest indicative of the Coastal Mountain Transition is a mixture of the lush temperature rainforests of the southeast and the spruce dominated forests of the interior, although they show more semblances to the boreal forest of the interior. The dominant disturbance regime in the maritime forests is wind throw. Extreme wind events in the region can cause the felling of large stands creating openings in the canopy and allowing soil churning to take place, which in turn provides new habitat for other trees (Forest Health Monitoring, 2004).

## Glaciers

Glaciers exist in several different forms in Alaska: ice fields, ice domes, mountain glaciers, valley glaciers, piedmont glaciers, cirque glaciers, hanging glaciers, and tidewater glaciers. Glaciers are primarily found in the south-central and south-eastern portion of the state, while a few are located in the Brooks Range to the north (Figure 7). Glaciers cover about 70,000 km$^2$ of the land surface of Alaska, which accounts for roughly 5% of the total area Arendt et al. (2002) identified 67 glaciers in Alaska and the surrounding Canadian Territories and separated them into seven geographic regions: Alaska Range, Brooks Range, Coast Range, Kenai Mountains, St. Elias Mountains, Western Chugach Range, and the Wrangell Mountains. Using this classification, there are 55 glaciers located solely in Alaska; 11 glaciers that cross the border between Alaska and Canada; and one glacier is located entirely within the Yukon Territory. If all of the glaciers in Alaska and the neighboring territories of Canada were considered, they would cover 90,000 km$^2$ and account for nearly 13% of the mountain glacier area worldwide (Dyurgerov and Meier, 2000; Meier, 1984).

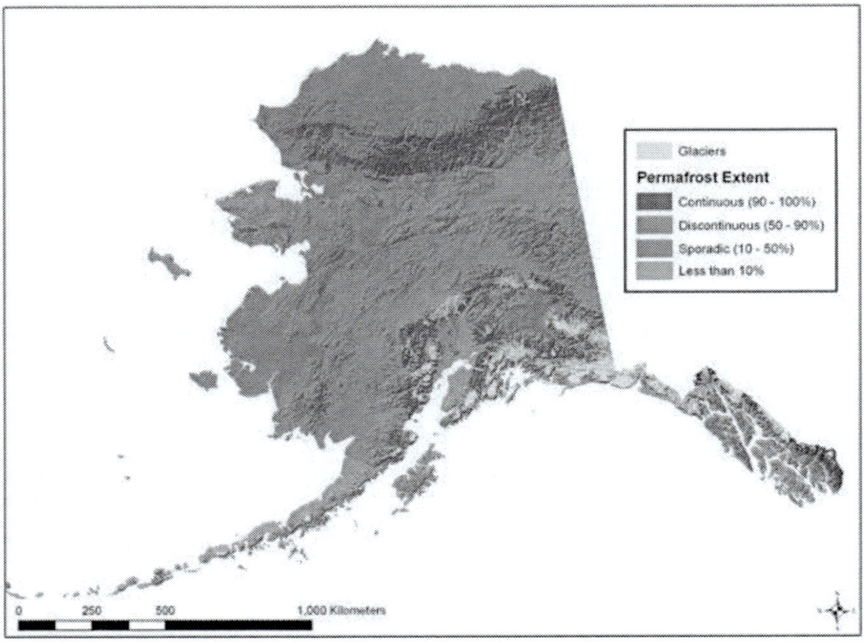


Figure 7. Two components of the cryosphere in Alaska: glaciers and permafrost. Glaciers cover roughly 5% of the land surface of Alaska. Continuous permafrost accounts for 33%, discontinuous 39%, and the sporadic permafrost zone 12% (data source: Brown et al., 1998).

## Permafrost

As defined by Muller (1947) permafrost is any thickness of soil or other surficial deposit which has been colder than 0°C for 2 years or more. In Alaska, permafrost varies in its depth, continuity, and ice content. There are four generalized categories of zones of permafrost: continuous (90 to 100% aerial extent), discontinuous (50 to 90%), sporadic (10 to 50%), and isolated (less than 10%).

About 85% of the land surface of Alaska is underlain by permafrost (Figure 7). Continuous permafrost is found beneath 33% of the land surface, discontinuous permafrost underlies 39%, and sporadic permafrost accounts for 12%. Continuous permafrost is mostly found in northern and western portions of the state. Discontinuous and sporadic permafrost are typically found in the interior and southern portion of the state north of the Maritime ecoregion.

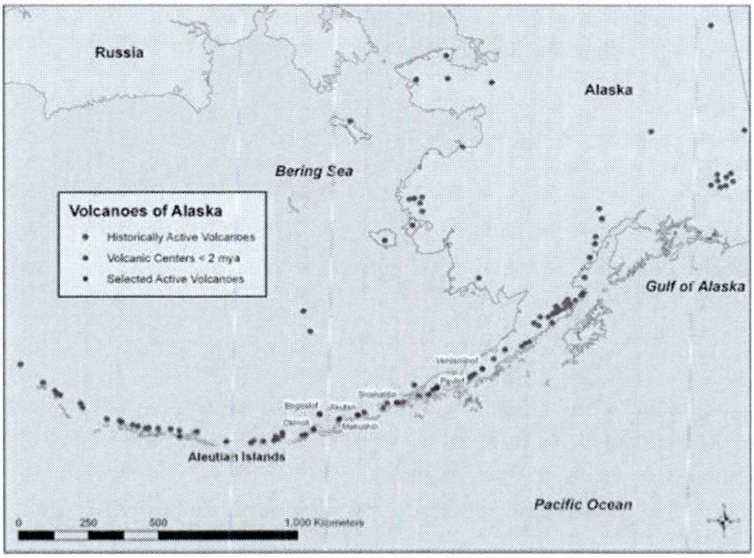


Figure 8. The volcanoes of Alaska. The volcanoes indicated in red are considered to be historically active volcanoes. The seven volcanes highlighted account for 60% of the eruptions over the historical period (data source: Cameron, 2005).

## Volcanoes

There are more volcanoes in Alaska than in the rest of the United States. Cameron (2005) provided the latitudes and longitudes of 127 volcanic centers in Alaska believed to be less than 2 million years old (Figure 8). Of these, more than 40 have been reported as having been a historically active volcano; that is having experienced a volcanic eruption, other than low level steam and gas venting, in historical time (Miller et. al., 1998; Cameron, 2005). The historic period in Alaska generally refers to the period since 1760, during which written records have been kept. At least 265 eruptions have occurred at the 29 most active volcanoes in historic times and seven volcanoes (Veniaminof, Pavlof, Shishaldin, Akutun, Makushin, Okmok, and Bogoslof Island) have accounted for 60% of the 265 eruptions. As a minimum estimate, there have been 1.1 to 1.3 eruptions per year since 1760; however, because much of Alaska remains to be sparsely populated, many eruptions may have gone unreported. A better estimate of eruptions began in 1945 as this marks the beginning of widespread air travel in the state. Since then, 90 eruptions have been reported from 23 different volcanoes. This results in a frequency of 1.8 eruptions per year (Miller et. al., 1998).

### Earthquakes

Alaska is located in a very seismically active area that is primarily the result of the collision between the Pacific Plate and the North American Plate. The denser Pacific Plate is moving northwestward towards southern Alaska, the Alaskan Peninsula, and the Aleutian Islands; where the two plates meet, the Pacific Plate plunges below the North American Plate creating a subduction zone. Much like the active volcano pattern, a great deal of the earthquake activity follows the Aleutian Arc, although there are also significant earthquakes that occur inland.

Since 1898 there have been nearly 600 earthquakes with a magnitude larger than 6.0 reported in and around Alaska and 5 earthquakes with a magnitude larger than 8.0 (Figure 9) (AEIC Earthquake Database, 2005). Within 160 km of Anchorage, there have been 18 earthquakes with a magnitude larger than 6.0 since 1898, the largest being the 9.2 magnitude quake in 1964. Within 160 km of Fairbanks, 12 earthquakes greater than 6.0 have occurred over the same time period. Of these, four have had a magnitude greater than 7.0. Conversely, there have only been 7 earthquakes, greater than 6.0, within 160 km of Juneau from 1898 to 31 May 2005.

## Socio-Economic Conditions

### General Population

The United States purchased Alaska from Russia in 1867 and since that time the trends in the settlement patterns and the growth of population throughout the state have been guided by boom and bust cycles (Error! Reference source not found.). The first such boom, the Klondike gold rush, occurred at the turn of the 20$^{th}$ Century. During the gold rush, the population doubled from roughly 32,000 people to 64,000 people, with peak gold production occurring in 1906.

In 1912 Alaska was organized as a territory, however, following the Klondike gold rush the population of Alaska declined between 1910 and 1920.

The 1920 Census reported the lowest population figure of the 20th Century at 55,000, but then gradually increased during the 1920's and 1930's. Some significant events that occurred during this time period consisted of the opening of the first pulp mill in Southeast Alaska (1922), the completion of the Alaska railroad (1923), and the establishment of farming families in the Matanuska Valley (1935).

Alaska's population began to increase rapidly in the 1940's coinciding with the establishment of the Fort Richardson Army Base and the Elmendorf Air Force Base. In the 20 years following their establishment the population had tripled from 72,500 to 226,000. Two other important events occurred over this 20 year period. During the 1950's, timber harvesting and processing boomed in Southeast Alaska. As a result, the first plywood mill opened in Juneau in 1953 followed by another pulp mill in Sitka in 1958. The end of this decade was marked by the proclamation of Alaskan statehood on January 3, 1959.

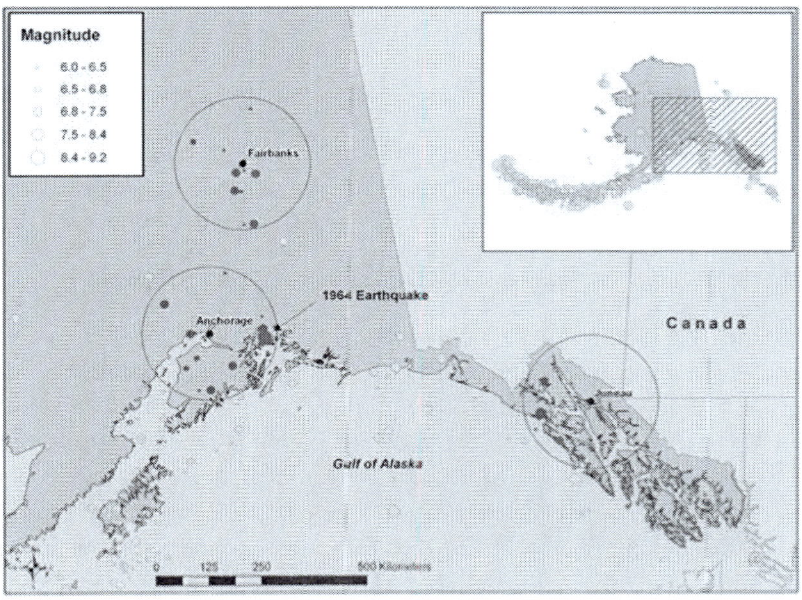


Figure 9. Earthquakes in Alaska with a magnitude greater than or equal to 6.0. The inset image in the corner shows all of the 6.0 or larger earthquakes. The zoomed image shows the earthquakes relative to prominent cities in Alaska (yellow indicates an earthquake of at least 6.0 magnitude and red indicates an earthquake of that severity within 160 km (100 miles) of either Anchorage, Fairbanks, or Juneau (data source: AEIC earthquake database).

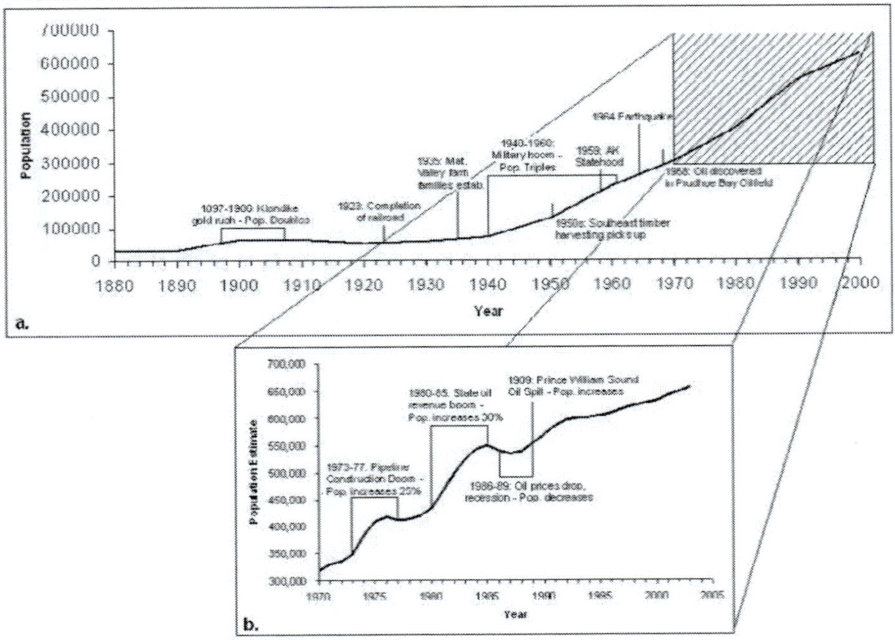


Figure 10. Alaska population, 1880 to 2004. (a) U.S Census Bureau population data from 1880 to 2000. The inset box (b) represents Alaska population estimates from 1970 to 2004. Estimates based on economic forces (data sources: Alaska Department of Labor and Workforce Analysis; Leask, 2001; and U.S. Census Bureau).

During the 1960's Alaska's population continued to gradually increase. At the end of the decade the Prudhoe Bay oil reserves, located in northern Alaska, were discovered and in 1969, a $900 million North Slope oil lease sale took place. This promise of new jobs resulted in a population boom during the mid-1970's and has been responsible for a great deal of social and economic changes since that time because a large share of the oil profits have gone into the state's economy (Leask et al., 2001). In 1973, construction began on the Trans-Alaska Pipeline and continued through 1977, during which time the population increased roughly 25%, from 348,000 to 412,000. Following the completion of the pipeline the population leveled off over the next three years.

The beginning of the 1980s marked some important events in Alaskan state history. In 1980, the Alaska legislature repealed the state income tax law and the legislature also established the Permanent Fund, setting aside one-fourth of all royalty oil revenue for future generations. Thus, the stagnant growth at the end of the 1970's was then followed by a major increase (30%)

in state population in the early 1980's. This coincided with a peak in state oil revenues. The Alaska National Interest Lands Conservation Act (ANILCA) passed in 1980 also added to the influx of people to Alaska at this time. ANILCA converted vast tracts of land in Alaska to national wildlife refuge lands, national wild and scenic river lands, the national forest lands, and national park lands which consequently led to an increase in the number of federal employees entering the state to undertake management responsibilities for these lands. Overall, the increase in population primarily occurred as a result of massive state spending; however, this boom ended in 1986 with the worldwide collapse in oil prices and Alaska experienced a recession and a decreasing population throughout the remainder of the 1980's (Leask et al., 2001).

Alaska's economy slowly recovered by 1990 and the state population had begun to grow again. Some of the growth that began to occur at the end of the 1980s and beginning of the 1990s was a result of the Exxon/Valdez oil spill, partially due to the massive workforce that was needed to cleanup the spill in Prince William Sound. The growth that has occurred since the early 1990s has been slow compared to that of the mid-1970s and the early to mid-1980s. The slower growth in population could be linked to the decline in North Slope oil production since the peak of the 1980's (Figure 10), which, in turn has corresponded to smaller state revenues (Leask et al., 2001). Currently, much of the growth is being driven by a population that has become more stable and one that shows a longer residence time than in previous decades. The number of residents staying in the state longer than 5 years grew by 20% from 1970 to 1990 (Leask et al., 2001).

The population of Alaska in 2004 was estimated to be 655,435 (US Census Bureau, 2005). According to the State of Alaska, Division of Community Advocacy the population is dispersed throughout 358 communities, with Anchorage being the largest urban setting with a population of 272,687 in 2004 (US Census Bureau, 2004). The state population is expected to gradually increase throughout the 21$^{st}$ Century as areas outside of Anchorage continue to be developed.

## Native Population

The Alaska Native and American Indian population accounts for more almost 20% of Alaska's statewide population (US Census Bureau, 2005). The native population has experienced many changes since the turn of the 20$^{th}$

Century. The native population declined dramatically following the arrival of Russian Explorers and Americans in the late 1700s and 1 800s, respectively, and continued through 1910. During this time several epidemics, such as, smallpox, measles, influenza, diphtheria, and tuberculosis devastated many of the native villages. After 1910 the native population gradually increased and during the 1950s public health in rural communities improved and the population began to increase, since 1960 the native population has almost tripled statewide and it is believed that the population reached the "pre-contact" level in the early 1970s. Much of this growth is attributed to a reduction in infant mortality which have declined by more than a third since the early 1990s. An improvement in health care for adults allowing life expectancy to increase also accounts for the increase in population. The average life expectancy among Alaskan natives was roughly 47 years in 1950, this increased to 64.4 in 1980, by 1990 it was 68.8, and in 1997 it was 69.5 (Goldsmith et al., 2004). There has also been a shift by Alaskan Natives from rural, subsistence lifestyles to urban living. From 1970 to 2000, the percentage of Alaska Natives living in urban areas increased from 17 to 32 percent (Leask et al., 2001).

## Jobs

As of 2004, there were 304,000 jobs statewide. This year marks the first time that the number of jobs in Alaska has reached the 300,000 mark. Historically, the number of jobs in Alaska surpassed 100,000 in 1972 and 200,000 jobs in 1982 (Erskine, 2005). Thus, it is evident the number of jobs in Alaska increased much more dramatically in the 1970s and early 1980s than it does today. Still, the historical increase in jobs statewide is impressive. Since 1961, there are five times as many jobs statewide and the state has posted an increase in jobs every year since 1988. Since 1960, state and trade industries created about half of the new jobs, state and local governments created about 20 percent, and resource and infrastructure industries added about one-third (Leask et al., 2001).

As mentioned before, job growth slowed in the 1990s and several of the industries in Alaska actually lost jobs from 1990 to 1999 (Figure 11). Since 1990, tourism has added more jobs than any of the other basic industries, while the biggest decline in jobs were from the military sector, the timber industry, and the petroleum industry.

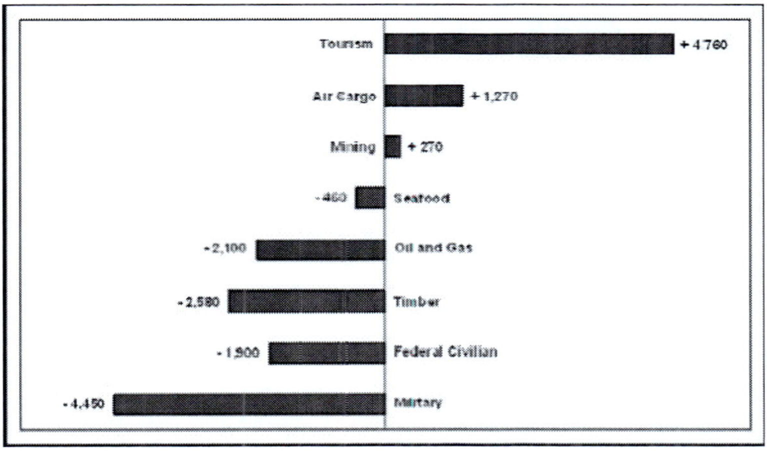


Figure 11. Graphic showing the gains (green) and losses (maroon) of basic industry jobs in Alaska from 1990-1999 (modified from Leask et al., 2001).

## Commodities

Commodities refer to tangible items that can be bought or sold. In Alaska, there are three major commodities that contribute to the economy: seafood, petroleum, and minerals. In 1965, seafood was the most valuable commodity; however, over the next 20 years this changed as oil and gas production surpassed seafood as the leading commodity, primarily a result of the development of resources on the North Slope. In 2000, seafood had dropped to the third most valuable commodity behind oil and gas and minerals. Minerals had become the second most valuable commodity in the state as the production of zinc increased, a result of the development of Red Dog Mine in northwestern Alaska, as well as a decrease in salmon prices worldwide due to pressure from foreign farmed salmon (Leask et al., 2001). Agriculture is also another commodity that contributes to Alaska's economy, however, it continues to remain small.

## Military

The United States Military holds a large presence in Alaska, both in terms of population and economics and has been this way since World War II. The state population tripled during the military boom between 1940 and 1960 coinciding with the establishment of the Fort Richardson Army Base and the Elmendorf Air Force Base in Anchorage. The number of military personnel in Alaska in 1960 was 33,000 making it the largest single employer statewide. In

recent times however, the military population has declined and by 2000 the population had dropped to 18,000 individuals. Military activities within the state however, still contribute about $7 of every $100 of gross state product (Leask et al., 2001).

## Tourism

Tourism has been one of the fastest growing contributors to the state's economy since the 1990s with annual visits increasing from roughly 1.1 million in 1994 to over 1.7 million in 2004 (Northern Economics, Inc., 2004). The increase in the number of tourists statewide is reflected in the number of jobs created in the tourism industry which added more jobs than any other basic industry through the 1990s (Leask et al., 2001).

A report prepared by Global Insight (2004), for the Alaska Department of Commerce provides an account of the importance of tourism to the state's economy. The report highlights the challenges of measuring travel and tourism, signifying that tourism is a part of many different industries and not one industry entirely. The report provided three different levels of measuring the contribution of travel and tourism: total sales, total economic contribution, and core industry. The latter measure of the travel and tourism industry provides the narrowest definition of the contribution of the tourism industry because it measures "only the direct impact of end-providers of goods and services to travelers" (Global insight). Therefore, this is the best measure to use for comparison with other industries. Thus, using core industry, in 2002, travel and tourism accounted for 3% of Alaska's gross state product, provided 26,159 jobs ($4^{th}$ overall and 9.1% of total employment), and provided $579 million in core labor income (benefits and salaries) to Alaska.

# Factors Affecting Land Surface Features in Alaska

The current status and trends of the land surface of Alaska is an amalgamation of many different factors. These factors can be broken down into two categories; natural forcing and anthropogenic forcing. Natural disturbances, such as forest fires, insect infestation, and coastal erosion can leave a marked impression on the land surface. In addition, a changing climate can lead to widespread changes in glaciers, permafrost and forests. Anthropogenic forces such as, changes in socio-economic conditions can also lead to localized as well as widespread surficial changes; exploration, urban development, and resource extraction can cause several noticeable changes in the land surface. Thus, in order to get an idea of the factors affecting land surface features in Alaska each of these features must be considered.

## Natural Drivers

The effects of a changing climate are presumed to be most rapid and dramatic in high-latitude areas (Serreze et al., 2000); thus, Alaska may be an important place to monitor these changes. Numerous studies have taken place within Alaska that have focused on changes in temperature and precipitation; frequency and severity of forest fires; the impact of insect infestation on the forests of Alaska; the size and extent glaciers; changes in permafrost; and changes in the coastline due to erosion. The effects of a changing climate have been well documented in Alaska and this section will serve as a synthesis of the basic trends that are apparent as a result of a changing climate.

### *Temperature and Precipitation*

Changes in climate occur at different spatial and temporal scales, therefore it is important to use caution when analyzing short and long term trends in climate data because they do not occur in a linear fashion; rather, they could occur as a result of a sudden shift in synoptic conditions (Hartmann and Wendler, 2005). Nevertheless, analyzing trends in climate data are an important factor to take into consideration when looking at changes in the land surface of a particular area.

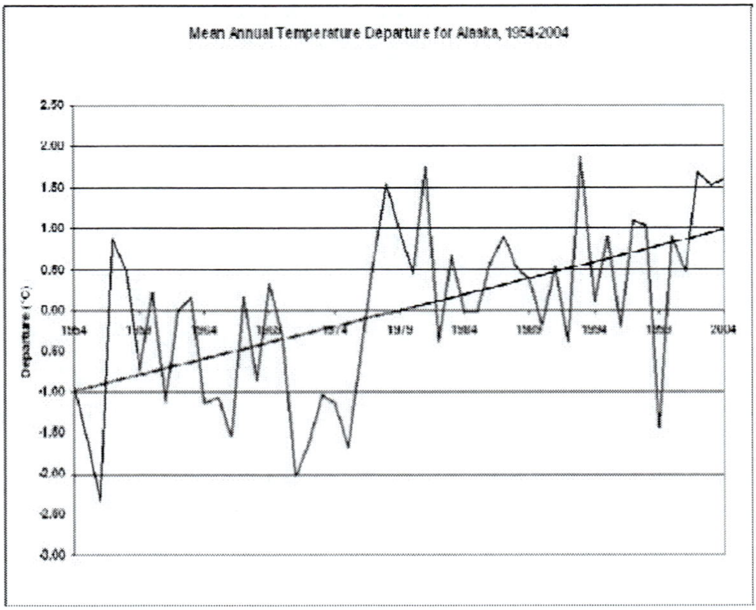


Figure 12. Graph showing the statewide mean annual temperature departure from 1954 to 2004; mean annual climate data has been plotted on the graph and a linear trend was run through the data to gauge the total change over the 50 year period (data source: Alaska Climate Research Center).

Temperature and precipitation data were collected from 20 of the 21 first-order weather stations in Alaska from the period 1954 to 2004; Valdez was omitted because the record only dated back to the 1970s (Alaska Climate Research Center). The mean monthly temperature and precipitation data were used to determine mean annual temperature and precipitation values as well as seasonal values. The weather station data indicate a mean temperature increase of 1.97°C from 1954 to 2004 (Figure 12). Using climatic regions similar to Shulski et al. (2005), regional patterns of temperature change emerge (Figure 13). The Arctic, Interior, West, and Southcentral regions have all experienced warming greater than 2.0°C,whereas the warming in the Southwest and Southeast regions has been less pronounced, 1.30°C and 1.52°C, respectively. Overall the majority of the warming has occurred in the fall and the winter (Table 1). Examining the graph closer it is evident that the changes in temperature over this 50 year period have not occurred in a linear fashion Around 1976, there was a shift in the Pacific Decadal Oscillation (PDO) from a negative phase to a positive phase (Mantua, 2001).

**Table 1. Seasonal temperature change from 1954 to 2004; 0 to 1in green, 1 to 2 in yellow, 2 to 3 in orange, and more than 4 degree increase in red. At several of the climate stations the warming has occurred during the winter and spring months. The seasons correspond to 3 month blocks; Spring = March, April, and May... (data source: Alaska Climate Research Center)**

| | Spring (°C) | Summer (°C) | Fall (°C) | Winter (°C) | | Annual (°C) |
|---|---|---|---|---|---|---|
| Anchorage | 1.97 | 1.10 | 1.06 | 4.02 | | 2.04 |
| Annette | 1.29 | 0.90 | 0.34 | 1.86 | | 1.10 |
| Barrow | 2.69 | 1.74 | 2.44 | 2.91 | | 2.45 |
| Bethel | 3.29 | 1.35 | 0.90 | 3.88 | | 2.35 |
| Bettles | 2.30 | 1.16 | 1.21 | 4.75 | | 2.36 |
| Big Delta | 2.29 | 0.76 | 0.99 | 4.76 | | 2.20 |
| Cold Bay | 1.49 | 1.22 | 0.69 | 0.98 | | 1.10 |
| Fairbanks | 2.55 | 1.17 | 0.79 | 4.16 | | 2.16 |
| Gulkana | 1.45 | 0.57 | 0.44 | 4.13 | | 1.65 |
| Homer | 2.20 | 1.93 | 1.45 | 3.52 | | 2.27 |
| Juneau | 1.90 | 1.55 | 1.08 | 3.18 | | 1.93 |
| King Salmon | 3.35 | 1.47 | 1.05 | 4.93 | | 2.70 |
| Kodiak | 1.27 | 0.94 | 0.34 | 0.69 | | 0.81 |
| Kotzebue | 1.51 | 1.30 | 1.82 | 3.41 | | 2.01 |
| McGrath | 2.84 | 1.61 | 1.29 | 3.76 | | 2.37 |
| Nome | 2.73 | 1.37 | 1.40 | 2.99 | | 2.12 |
| St Paul | 2.11 | 1.76 | 1.01 | 1.13 | | 1.50 |
| Talkeetna | 3.02 | 1.86 | 1.89 | 4.61 | | 2.84 |
| Yakutat | 1.87 | 1.05 | 0.56 | 2.61 | | 1.52 |
| | | | | | | |
| Statewide Change | 2.22 | 1.31 | 1.09 | 3.28 | | 1.97 |

The PDO primarily affects the North Pacific and has been described as a long-lived El Nino-like pattern of Pacific climate variability; while PDO phases are believed to persist for 20 to 30 years (Mantua, 2001). This is represented by the spike in mean annual temperature around the middle of the graph. Since 1976, there has actually been little warming statewide, thus the majority of the warming that has taken place can be accounted for by this shift to a positive phase of the PDO. Hartmann and Wendler (2005) have reported that mean annual and seasonal temperatures have been as much as 3.1°C higher during this positive phase of the PDO when compared to the negative phase. The authors also reported a deepening of the Aleutian low in winter and spring which has lead to increased advection of relatively warm and moist air from the south; therefore, this could explain why much of the warming has taken place during the winter and spring.

Analyzing the precipitation data shows that there has been an increase in precipitation of about 1 cm throughout most of the state. The precipitation data however, do not show the strong seasonal pattern that is evident in the temperature record, nor does it show a strong relationship with the shift in the PDO in 1976. The only region to show a decrease in precipitation is the Arctic region (but this is only represented by one site) (Figure 14). Curtis et al. (1998) also found that the Arctic region has experienced a decrease in precipitation. The authors determined that some causes of this decrease could be a result of wind direction changes, decreasing variability in surface pressure, and decreasing cloudiness. Overall, the precipitation data is believed to be somewhat problematic due to precipitation catchment problems and differing observation techniques (Shulski et al., 2005). Other recent studies that have analyzed temperature and precipitation data for the state of Alaska have come to similar conclusions indicating that both temperature and precipitation have increased statewide over the last half of the 20$^{th}$ Century (Stafford et al., 2000; Shulski et al., 2005).

### *Fires*

Forest fires are a major part of the ecosystem processes in Alaska and are responsible for wide scale changes in the land surface. On average, up to ten times more forest area is burned on an annual basis in Alaska when compared to the rest of the United States (Court and Griffith, 1992). This is largely a function of the size of Alaska, its immense boreal forest region, and low population densities which translate into allowing more fires to burn unimpeded. During the 2005 fire season, there were 624 fires throughout the state that burned an area of 18,650 km$^2$. This was the second highest total of area burned from the period 1990 to present and the third highest total since 1950.

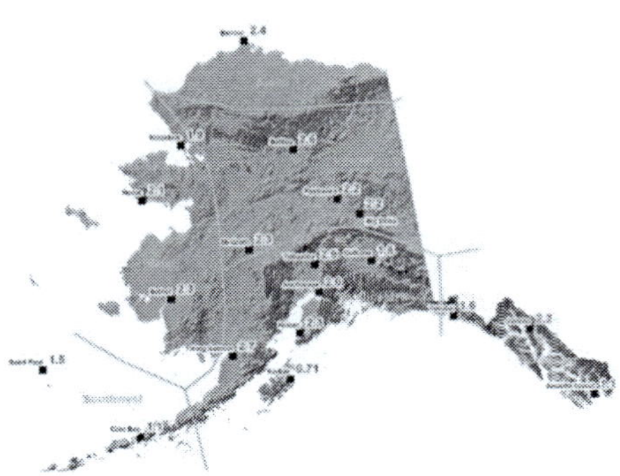

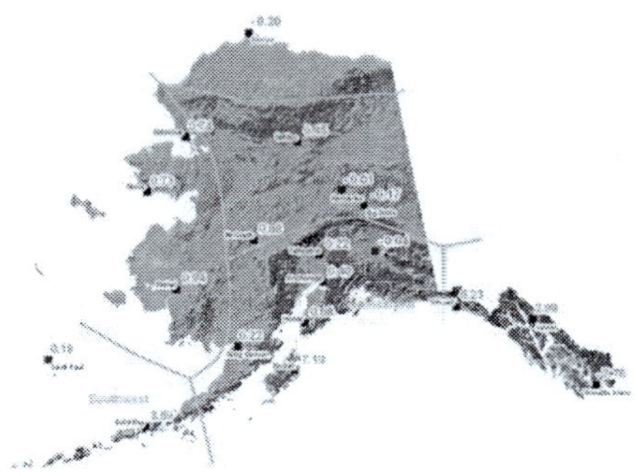

Figure 13 (left). Climate divisions in Alaska based upon Shulski et al. (2005). The greatest changes in temperature have occurred in the arctic and interior regions, while the least amount of change has occurred in the southeast and southwest regions.

Figure 14 (right). Mean annual precipitation change in Alaska, 1954 to 2004; blue indicates decreasing values and green indicates increasing values. Climate divisions based on Shulski et al. (2005).

The largest fire season on record in Alaska was the 2004 fire season, surpassing the 20,230 km$^2$ burned in 1957 with a total acreage burned of 27,200 km$^2$. In 2004, there were 706 fires across the state and of these 706 fires, 424 were caused by humans and 259 were caused by lightning strikes. The record breaking fire season of 2004 was initially predicted to be an "average" fire season. The winter snow pack throughout the state was 70% to 120% of the normal average, during the late spring/early summer intense convection and record-breaking rainfall occurred in interior Alaska, and there was only 103 km$^2$ burned by the middle of June (Alaska Fire Service, 2004). However, conditions began to shift during the middle of June and the resultant fire season can be attributed to the following factors. Following the record rain fall totals for the interior during the early summer, drought conditions set in. For the month of June, precipitation for the region was 25 to 50 % of normal and the temperatures were 6°F to 10°F above normal. During July, the temperatures were only slightly above normal and in August the same drought conditions returned to the interior. Fairbanks experienced the least amount of precipitation in the month of August in over 100 years of records. Thus, the persistent drought conditions throughout the interior during the summer of 2004, the above normal temperatures from May to August, the below normal precipitation from June to August, and the abundance of lightning strikes from the end of May to the middle of July contributed to the extreme fire season of 2004 (Richmond and Shy, 2005).

Fire reporting datasets date back to the 1950's. The BLM – Alaska Fire Service has compiled datasets for the time period 1950 to 2004. The data are available as a shapefile which allow for limited spatial analysis because of the coarseness of the data and probability of unrecorded fires. The originators of the data warn that there are at least 300 missing fires in the dataset and that this will impact the results of its use; however, some basic acreage burned estimates and other measures can be made. Fires recorded from the period 1950 to 1987 exceeded 4 $km^2$ in area burned and fires recorded from 1988 to 2004 consisted of fires greater than 0.4 $km^2$. Thus, it appears that more fires have been reported since 1988 simply based on the change in fire reporting status. Nonetheless, from 1950 to 2004, there have been 1,950 fires reported which account for a burn area of 188,900 $km^2$ across the state of Alaska. Of the 1,950 fires, about 25% are between 0.4 $km^2$ and 4 $km^2$ indicating that they were post-1987; however, these smaller fires only account for an addition of roughly 0.4% of the area burned between 1950 and 2004. Of, the majority of the fires that have occurred in the interior region of the state, roughly 93% of the fires burned have occurred in the Intermontane Boreal region (Figure 15).

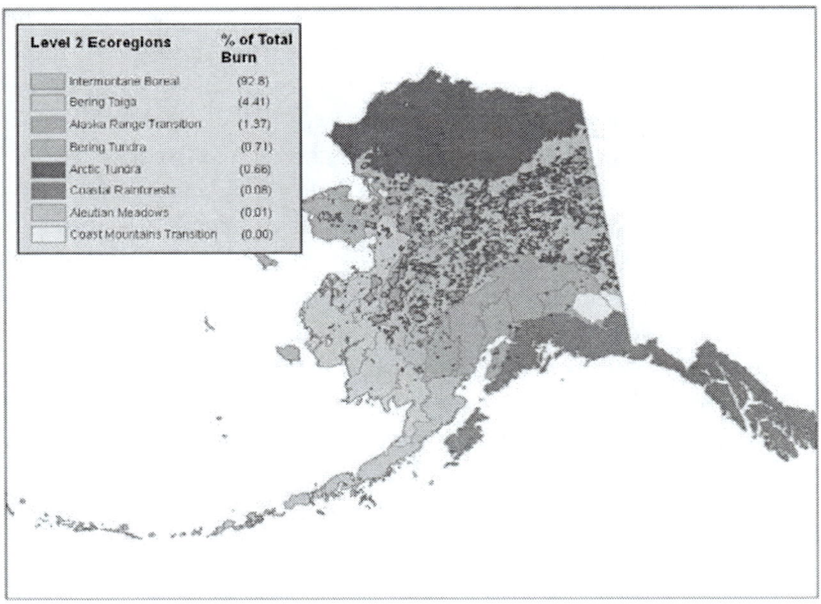


Figure 15. Fires in Alaska from 1950 to 2004. The majority (92.8%) of the fires have occurred in the Intermontane Boreal region (data source: BLM – Alaska Fire Service, Nowacki et al., 2001).

More detailed statewide fire statistics have been compiled by the Alaska DNR and the Bureau of Land Management since 1990 (Figure 16). These data indicate that since 1990, there have been 9,343 fires observed and 100,100 km$^2$ burned throughout the state. The 16 year average (90-05) for the state is 585 fires per year with a total of roughly 6,300 km$^2$ burned. The last two fire seasons, 2004 and 2005, have been the most active fire seasons during this time period. They have exceeded the 16 year average by a factor of nearly 4.5 and 3, respectively.

***Lightning and Fire***

The BLM has operated a cloud-to-ground lightning strike automated network since 1976, with 9 stations in Alaska and 3 in the Yukon Territory. The location of a lightning strike is estimated by triangulation, thus there has to be at least two sensors that detect the strike. The location of a detected lightning strike is assumed to have a positional accuracy of roughly 2-4 km and a detection efficiency of 60% - 80%, however, a study to test the detection efficiency and positional accuracy have not been conducted in Alaska (see Dissing and Verbyla, 2003).

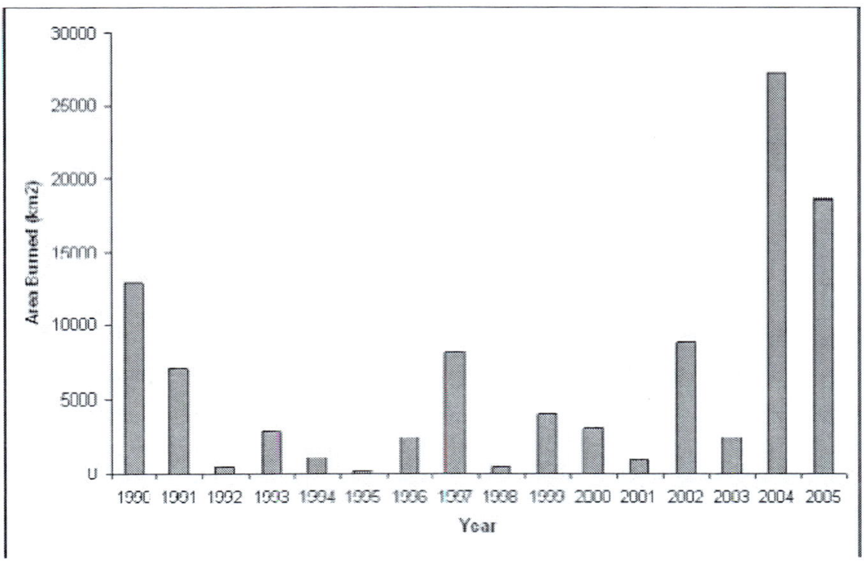


Figure 16. Total forest area burned (km2) in Alaska from 1990 to 2005 (data source: BLM – Alaska Fire Service).

Utilizing the lightning strike datasets from the Bureau of Land Management – Alaska Fire Service, recent studies have been conducted on the relationship between cloud-to-ground lightning strikes and wildfires in Alaska (McGuiney et al., 2005; Dissing and Verbyla, 2003). Dissing and Verbyla (2003) compared the spatial patterns of lightning strikes in the interior and their relationship with elevation and vegetation between 1986 and 1999. The McGuiney et al. (2005) study looked at the relationship between lightning strikes and fires as a whole between 1986 and 2002 and focused on distinguishing between human caused fires and lightning caused fires.

Dissing and Verbyla (2003) used the lightning strike dataset to determine where lightning strikes preferentially occurred. At the regional scale (interior Alaska), the authors found that the number of lightning strikes were driven by elevation and percent cover of boreal forest and location. These factors combined to explain 66% of the variation in lightning strike densities in the interior. Thus, lightning strikes appear to preferentially occur in continental locations between 1000 m and 1200 m in areas with around 50% to 70% boreal forest coverage.

McGuiney et al. (2005) utilized the lightning strike dataset in order to relate lightning climatology and wildfires in the state over the period 1986 to 2002. Over this time period, there has been an average of 32,400 lightning strikes per year and approximately 90 days per year in which lightning occurs. This study reported that almost 100% of the lightning strikes occur over the months May through August and nearly 91% of the lightning strikes occur in June and July and were a result of thunderstorm activity and hypothesized to be driven by synoptic conditions or convective lifting. The authors reported that the majority of the lightning strikes in the state were a result of convective lifting because there is a lag in the maximum stroke count of about 3 hours following solar noon. The authors also found a correlation between daily maximum temperature and the number of lightning strikes in this region of Alaska. For thunderstorm activity to take place, the temperature must be higher than 10°C, and for a thunderstorm associated with an extreme lightning event (>2000 strikes) the temperature must be greater than 21°C (McGuiney et. al. 2005).

Although humans have been the cause of more than two-thirds of the forest fires since 1986, lightning strikes are responsible for the majority of the area burned in the state. Lightning strike fires have accounted for almost 93% of the burned area each year from 1986 to 2002. McGuiney et al. (2005) found a positive correlation between lightning strikes and wildfire starts between 2000 and 2002 ($R^2$ = 0.45), a negative correlation between precipitation and wildfire starts between 2000 and 2002 ($R^2 = 0.21$) (McGuiney et. al., 2005). Thus, with potentially warmer conditions occurring in Alaska, the number of lightning strikes may increase as a result of increased convective lifting.

***Insect Infestation***

The impact of insect infestations on Alaskan forests as a disturbance agent is believed to be linked to a warming climate (Juday et al., 1998). A climate shift in which there are higher temperatures and less precipitation allows insect reproductive rates to increase, as well as, weakening the strength of the forest by inducing stress from drought (Forest Health Monitoring, 2004). The primary insects that have contributed to disturbances throughout Alaska can be divided into two categories: bark beetles and defoliators. Bark beetles are considered to be more of a threat to Alaska's forests than the defoliators because the bark beetles cause a greater incidence of tree mortality (Forest Health Monitoring, 2004). Defoliators attack the leaves and needles of trees which lead to increased stress but not necessarily tree mortality (USFS-Forest Health Conditions in Alaska, 2004).

The dominant species of bark beetle that has lead to widespread disturbance is the Spruce bark beetle (*Dendroctonus rufipennis* Kirby), while the Western Balsam Bark Beetle (*Dryocoetes confuses* Swaine), the Eastern Larch Beetle (*Dendroctonus simplex* LeC.), and the Engravers (*Ips perturbatus* Eichhoff) have also contributed to forest disturbance. Many of the disturbances caused by bark beetles lead to changes in the structure and composition of the forest ecosystem as well as alter the socioeconomic benefits of the forest. Some examples are loss of harvestable timber, long-term stand conversion, changing wildlife habitats and behavior, impacts on aesthetic value, increases in fire hazard conditions, and impacts on fisheries and watersheds (Forest Health Monitoring, 2004). A predominant spruce bark beetle attack began in the late 1980s throughout the state with some areas having been impacted more than others (USFS-Forest Health Conditions in Alaska, 2004). The beetles bore into the bark of the spruce trees and feed on the thin phloem layer between the bark and the wood. This essentially starves the tree through lack of nutrients and causes mortality. Since 1989 there have been 13,000 $km^2$ of the forest impacted statewide. The two areas that have received the largest disturbance from the spruce bark beetles have been the Kenai Peninsula and the Copper River region.

The outbreak on the Kenai Peninsula began in 1989 and since that time over 5,600 $km^2$ of forest has been lost to spruce bark beetle infestations. This has been the largest case of tree mortality due to a single insect outbreak in all of North America (Werner, 1996). The affected tree species consisted of white spruce ((*Picea glauca* (Moench) Voss) and Lutz Spruce (*Picea xlutzii* Little) The peak of the outbreak occurred in 1996 and in recent years the impacts of the spruce bark beetles on the Kenai Peninsula has waned primarily because the bark beetles have destroyed the majority of their food source (Figure 17).

The Copper River Basin has also been heavily impacted by the spruce bark beetle with roughly 2,000 km$^2$ of forest lost to the infestation from 1989 to 1998 (USFS-Alaska State and Private Forestry Aerial Insect and Disease Surveys, 1989-1998). Wesser et al. (2001) found that within the Wrangell-St. Elias Natioal Park and Preserve region of the Copper River Basin over 800 km$^2$ had been impacted over the same time period (Error! Reference source not found.8). The spruce bark beetle infestation primarily occurred in white spruce (*Picea glauca* (Moench) Voss) forests within the park.

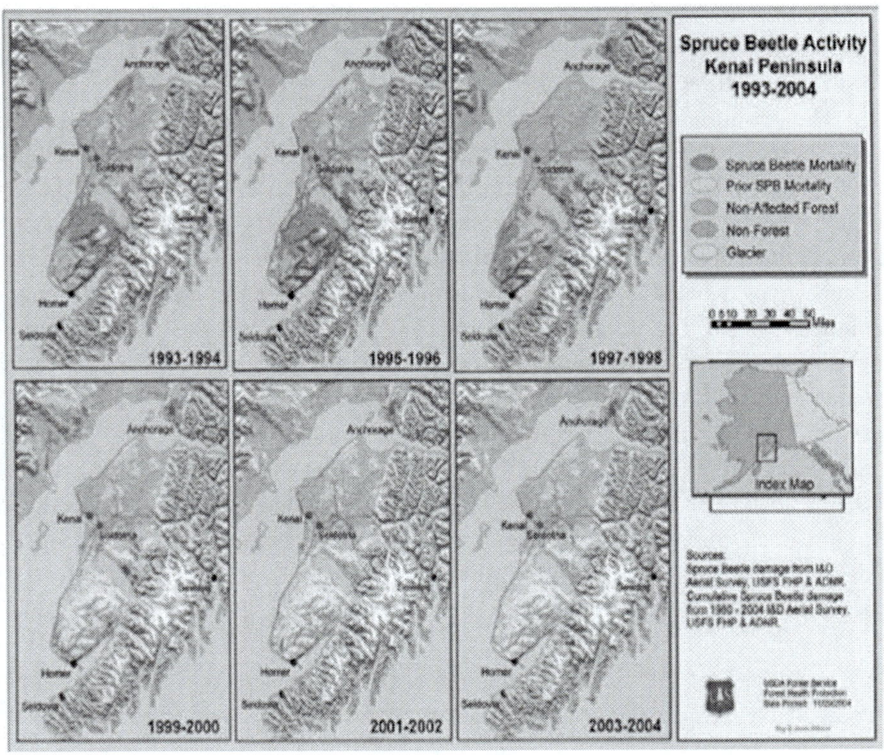


Figure 17. Spruce bark beetle impacts on the Kenai Peninsula from 1993 – 2004. The outbreak peaked in 1996, yet there are still newly impacted areas each year (Source: Forest Health Conditions in Alaska – 2004).

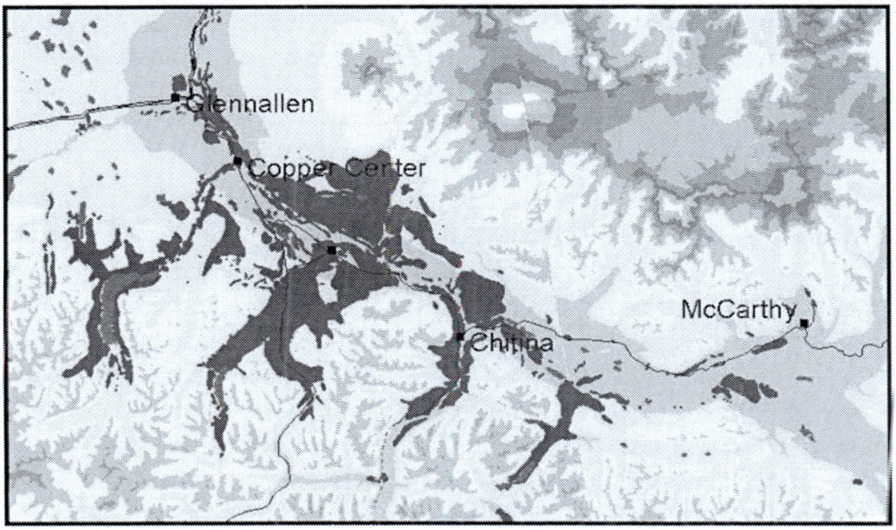


Figure 18. Areas within the Copper River region that have been impacted by the spruce bark beetle (red) since 1989 (Source: Wesser and Allen, 2001).

In addition to bark beetle kills, the defoliators found within Alaska are a major concern and consist of the spruce aphid (*Elatobium abietinum* Walker), spruce budworm (*Choristoneura fumiferana* Clemens), western black- headed budworm (*Acleris gloverana* Walsingham), yellow-headed spruce sawfly (*Pikonema alaskensis* Rohwer), hemlock sawfly (*Neodiprion tsugae* Middleton), larch sawfly (*Pristiphora erichsonii* Hartig), aspen leaf miner (*Phyllocnistis populiella* Chambers), birch leaf roller (*Epinotia solandriana* L.), willow leaf blotch miner (Micrurapteryx salicifolliela *Chambers*), and the noctuid defoliator (*Sunira verberata* Smith) (Forest Health Monitoring, 2004). The defoliator outbreaks tend to be cyclic in nature and related closely to climatic conditions (USFS-Forest Health Conditions in Alaska, 2004). Higher temperatures generally increase the adult emergence and survival rate and milder winter temperatures aid in the over wintering survival of adults. Of these species, the spruce aphids, spruce budworms, western black-headed budworms, larch sawflies, and aspen leaf miner have had the most impact on Alaskan forests (USFS-Forest Health Conditions in Alaska, 2004). Even though defoliators are found in all forests of Alaska, there effect on the maritime forests of Alaska appears to be more widespread over space and time (USFS-Forest Health Conditions in Alaska, 2004). The impacts of a defoliator outbreak can effect the feeding behavior of wildlife in the effected forest by altering the nutrient cycling of the ecosystem. This in turn may stunt the growth of trees, consequently decreasing economic returns, as well as altering the aesthetic value of the forest (Forest Health Monitoring, 2004).

### *Changes in Glaciers*

The glaciers of Alaska are believed to be changing rather rapidly in response to modern regional climate changes. Roughly 5% of the land surface of Alaska is covered by glaciers. The USGS has been monitoring two glaciers in Alaska, Gulkana and Wolverine, to analyze changes in climate, glacier geometry, glacier mass balance, glacier motion and stream runoff, since the 1960's. Gulkana Glacier (63°14'N) is located in the Alaskan Range and Wolverine Glacier (60°23 'N) is located in the Coastal Mountain Range in the Gulf of Alaska. Both glaciers have shown a drastic reduction in cumulative net balance since the late-1980's (Figure 19) and are indicative of changes taking place in the glaciers throughout the state and the neighboring Canadian territories (Arendt et. al., 2002). Arendt et al. (2002) used airborne laser altimetry to study the volume changes of 67 glaciers in Alaska and Canada from the mid-1950s to the mid-1990s representing roughly 20% of the 90,000 $km^2$ glaciated area. The overall results show a dramatic increase in glacier thinning however, from the mid-1990's to 2001 the rate of thinning was -1.8 m/year, while the rate of thinning was only -0.7 m/yr from the mid-1950s to the mid-1990s. Thus, the 5 to 7 year recent period has exhibited a rate of glacier thinning more than twice as fast as the 40 year early period. Extrapolating the results to include all of the glaciers in Alaska and neighboring Canada, the estimated annual change in volume is -52 +/- 15 $km^3$/yr.

Arendt et al. (2002) believe that this loss of volume has contributed to rising sea levels and that the loss of glacier area in Alaska's mountain glaciers have contributed to roughly 9% increase in sea level over the past 50 years. Thus, over the past 50 years it is believed that Alaska's melting glaciers have made the largest glacial contribution to rising sea levels worldwide. Even though the authors suggest that the rates of thinning could be related to a warming climate, they indicate that in some cases, the rapid thickness changes of surge-type glaciers probably occur independently of climate fluctuations (Arendt et. al., 2002).

### *Other Examples of Glacier Change*

Overall, land terminating valley and mountain glaciers are behaving differently than their tidewater counterparts (Arendt et. al., 2002). There are 8 calving glaciers in the state that are advancing and gaining mass. Each of these tidewater glaciers is believed to be linked more to the interaction between fjord depth, ice thickness, and calving rate, while climate change plays a secondary role (Meier and Post, 1987). Thus, changes in glaciers occurring as a result of a changing climate are better reflected in non-tidewater ice masses.

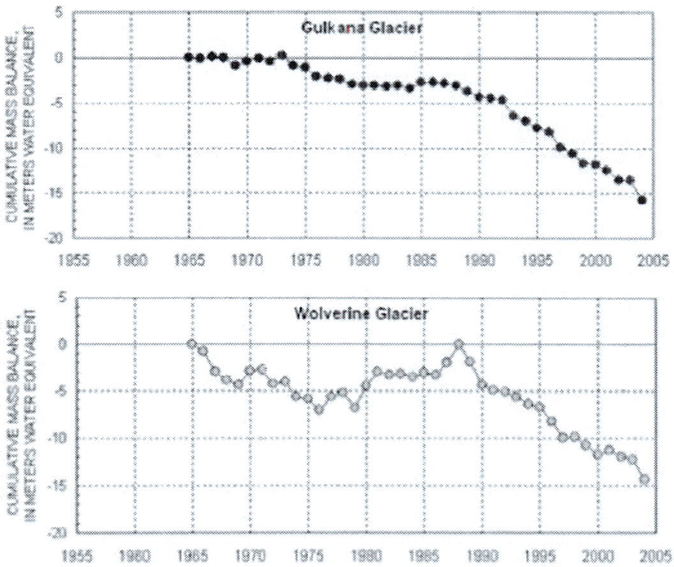


Figure 19. Changes in cumulative mass balance for Gulkana (Alaska Range) and Wolverine glaciers (Coastal Mountain Range) from 1965 to 2004 (Source: USGS, Benchmark Glaciers).

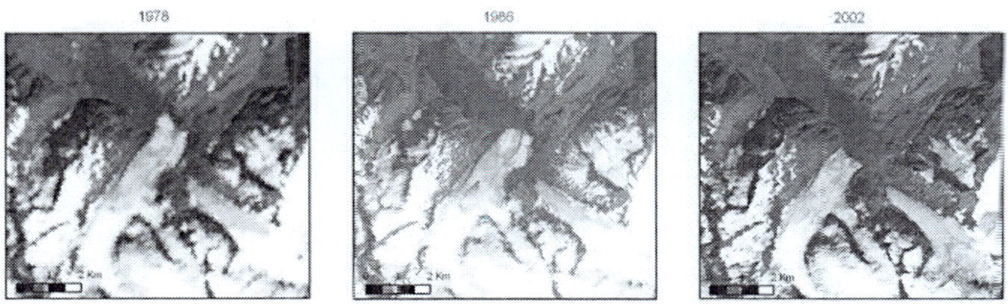


Figure 20. Portage glacier is located in center of each of these frames. The first frame shows the glacier in 1978. The next two frames show the glacier in 1986 (the year that the visitor center was constructed) and 2002, respectively. Over this time period the glacier has receded by such an extent that it can no longer be seen from the visitor center.

### *Glaciers in Central and Southcentral Alaska*

Portage Glacier is one of Alaska's most visited tourist attractions, located about an hour's drive south of Anchorage. The Begich – Boggs Visitor Center was constructed in 1986 and at that time the snout of the glacier was visible from the visitor center. Since that time however, the glacier has retreated and by 1993 it could no longer be seen from the visitor center (Figure 20). Much of the retreat taking place at Portage Glacier has been attributed to erosion from the warming of the proglacial lake (Blanchet, 2002). Exit Glacier is another one of Alaska's most visited glaciers. It emanates from the Harding Icefield and is a land terminating glacier located near Seward, Alaska. This glacier is also showing a pattern of retreat. Cusick (2001) determined that Exit Glacier reached its most recent maximum extent in 1815. Since that time the glacier has been receding in a series of waves, with some years of intensive retreat followed by years of slow to moderate retreat.

### *Glaciers in the Brooks Range*

The McCall Glacier is a land terminating glacier located in the Brooks Range inside the Arctic National Wildlife Refuge. It is believed that this glacier began its retreat towards the end of the 1 800s with monitoring beginning in the late-1950s; photos taken in 1958 and 2003 by Post and Nolan, respectively, offer a comparison of the glacier between the two time periods (Figure 21). However, due to the remoteness of the glaciers in the Brooks Range, this region remains understudied.

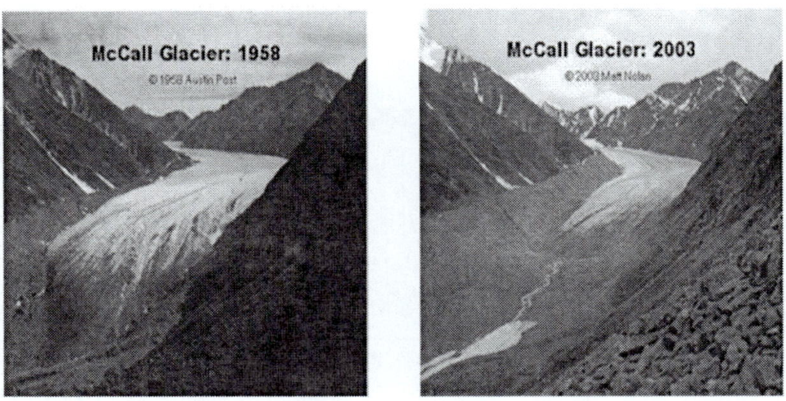


Figure 21. Historic and recent oblique photographs of McCall Glacier, Brooks Range, Alaska. The photo on the left was taken in 1958 and the photo on the right was taken at the same spot in 2003 (Source: Nolan, M., UAF).

***Glaciers in Southeast Alaska and Prince William Sound***

Glacier Bay, Alaska has only formed in recent times (the last 200 years) following a rapid retreat of glaciers in the region and is currently composed of both tidewater and non-tidewater glaciers (Seramur et al., 1996). For the most part these glaciers have been retreating. The Muir Glacier, a tidewater glacier, has been in a fairly steady state of retreat since the late 1970s. Landsat derived measurements have revealed that the glacier retreated 0.73 km per year between 1973 and 1982, and 0.04 km per year between 1982 to 1992 (Hall, 2005). Six other glaciers within the bay have retreated more than 1 km from 1966 to1986, while some have advanced slightly, and one, the Carroll Glacier, a tidewater glacier, has advanced more than 5 km (Hall et al., 1995).

The Shoup Glacier is a tidewater glacier located in northeast Prince William Sound that began its documented retreat in 1750 and around 1850, the glacier became somewhat stable on a bedrock shoal, which was followed by minor advances and retreats for about 100 years. In 1957, the glacier receded off of the shoal into deeper water and since that time the retreat is believed to be controlled by water depth (Brown et al., 1982). Between 1957 and 1990 the glacier retreated about 2.0 km exposing a deep tidal basin (Post and Viens, 1994).

Other examples of tidewater glaciers in Alaska are the Hubbard Glacier, near Russell Fjord, and the Harvard and Yale Glaciers of College Fjord, northwest Prince William Sound. The Hubbard Glacier has advanced about 24 meters per year from 1895 to 2001 and in 1986 and again in 2002 actually blocked Russell Fjord creating the two largest glacial-lake outburst floods in historical time (USGS fact sheet 001-03). The Harvard and Yale Glaciers are the two largest tidewater glaciers in College Fjord. The Harvard Glacier has been advancing since at least 1905 and since 1931 it has advanced approximately 20 meters per year (Sturm et al., 1991). The Yale Glacier on the other hand has retreated about 50 meters per year over the same time period (Sturm et al., 1991). The behavior of these two adjacent glaciers suggests that their terminus behavior is being controlled more by the complex interaction of factors in the tidewater environment than by climate change (Sturm et al., 1991).

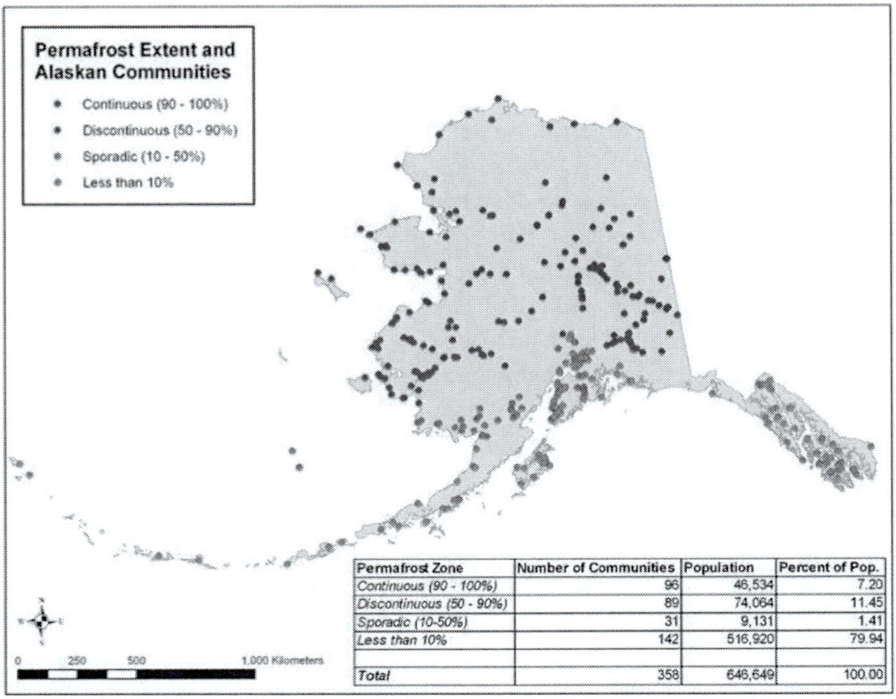


| Permafrost Zone | Number of Communities | Population | Percent of Pop. |
|---|---|---|---|
| *Continuous (90 - 100%)* | 96 | 46,534 | 7.20 |
| *Discontinuous (50 - 90%)* | 89 | 74,064 | 11.45 |
| *Sporadic (10-50%)* | 31 | 9,131 | 1.41 |
| *Less than 10%* | 142 | 516,920 | 79.94 |
| | | | |
| *Total* | 358 | 646,649 | 100.00 |

Figure 22. The communities in Alaska as they relate to the permafrost zones found within Alaska. More than 20% of the population in state live in areas with more than 10% coverage by permafrost (data sources: Alaska Community Database; Brown et al., 1997,1998).

## *Changes in Permafrost*

For the most part, air temperature controls the existence and condition of permafrost but other factors such as substrate and groundwater contribute as well. A warming climate could potentially create widespread changes in permafrost across Alaska. As mentioned previously, roughly 85% of the state is underlain by permafrost and degradation of permafrost throughout the state could impact the socioeconomic conditions statewide because much of the current infrastructure resides atop areas underlain by permafrost (Figure 22). In addition, almost 130,000 people (~20%) live in areas that could be adversely affected by permafrost degradation and significant stretches of roadways traverse these populated regions.

There have been recent studies that have alluded to the warming of surface permafrost temperatures across Alaska (Osterkamp and Romanovsky, 1999; Jorgenson et al., 2001). Since the 1940's the USGS has monitored permafrost temperatures in northern Alaska suggesting that the temperature of the upper permafrost has warmed by 2-4°C through the mid-1980's (Lachenbruch and Marshall, 1986). Other data suggest that the permafrost surface began to warm extensively around the end of the 1980's (Clow, pers. comm.). A study by the USGS on the Arctic Coast Plain near Teshekpuk Lake has found that there has been a 3.6°C increase in surface temperatures since 1989. Other studies have also noted similar increases in near surface permafrost temperatures near Prudhoe Bay and the Arctic Ocean on the order of 3-4°C, a warming of about 2°C over the rest of Arctic Coastal Plain, and typically 0.5°C-1 .5°C in the zone of discontinuous permafrost has also been found (Nelson et al., 2003). Osterkamp and Romanovsky (1999) reported that the temperature of the permafrost had warmed at all of their study sites along a north – south transect from Prudhoe Bay to Glennallen through 1980's and 1990's

Widespread warming of the upper permafrost can have a large impact on the communities and infrastructure of Alaska. This is primarily due to the ice content of the permafrost. It is the thawing of the permafrost under ice-rich conditions that has the largest potential impact on society. Thermokarst (ground subsidence) will develop under such conditions which can affect buildings and roads as well as natural habitats.

### *Coastal Erosion*

Nearly 85% of the population of Alaska lives in an area considered to be a part of the coastal zone (Bernd- Cohen and Gordon, 1999). In a study conducted by a state task force in 1984, 62 communities were identified with known erosion problems (State of the Beach, 2005). The coastline of Alaska is also an important economic component of the state. A large proportion of the profitable industry (commercial fishing, oil and gas development, logging, and tourism) has its roots along the coasts of Alaska. Assuming an estimated 76,000 km long coastline, roughly 70% (55,000 km) is considered tidal shoreline, and only about 1% of the coastline of Alaska is considered to be critically eroding. (Bernd-Cohen and Gordon, 1999).

Coastal erosion is driven by the composition of bluff sediments in the coastline and their interaction with rising sea levels, storm events, and in some locations earthquakes and tsunamis. Coastal erosion has less of an impact in areas with rocky shorelines than in those areas with unconsolidated sediments and permafrost. Rising sea levels and storm surges have affected large areas of coastline in the Bering, Arctic, and Cook Inlet regions whereas, earthquakes have had more of an impact in the regions along the Gulf of Alaska and southeast.

Mason et al. (1997) identified coastal communities in which bluff erosion was a major hazard and all of the communities occurred in the Arctic, Bering, or Cook Inlet region (Error! Reference source not found.). Much of the coastline of the Bering and Arctic regions consist of ice-laden, unconsolidated sediments that are vulnerable to coastal flooding and bluff erosion and collapse, whereas, much of the sediments in the Cook Inlet region are composed of glacial outwash composed of sands and gravels. In contrast, the coastline of the Aleutian Chain, Gulf of Alaska, and southeast Alaska consists of bedrock bluff or cliffs which are resistant to erosion over the same time span. The driver of coastal erosion in this region is typically a result of earthquake activity and block collapse. Earthquakes can also cause massive uplift of the land surface which can aid in the protection of the shorelines in this region although, associated tsunamis may have a large potential for shoreline destruction.

As an example of the effect of earthquakes on coastal stability, the magnitude 9.2 earthquake of 1964 that impacted southern Alaska caused the land surface within the vicinity of Prince William Sound to tilt, trending from the southern end of Kodiak Island northeastward to Prince William Sound overall uplift of the land occurred. On the southwestern end of Montague Island in Prince William Sound a vertical displacement of about 13-15 meters occurred. Conversely, over much of the Kenai Peninsula and the north and west portion of Prince William Sound the land surface subsided, resulting in an increase in coastal erosion along the western Kenai Peninsula (Stover and Coffman, 1993).

In Kachemak Bay in 2003 a beach monitoring camera station that observes coastal processes was installed by the USGS on the north side of the bay in Homer. Two of the primary objectives of this project are to monitor coastal erosion and loss of near shore habitat. In an area adjacent to the camera, Bishop's Beach, coastal erosion averages 0.5 – 1.0 meter per year. Bluff erosion on the north side of Kachemak Bay is primarily the result of storms that come from the southwest at high tide. Many structures have been lost to date and erosion has actually enhanced as some shoreline armor has been constructed. The construction of shoreline armor reduces erosion at the particular spot where it is constructed, however, it increases erosion in adjacent areas (Mason et. al., 1997).

### *Example of Coastal Erosion: Chukchi Sea Coast and Elson Lagoon*

Barrow is the largest town on the North Slope and the largest town that far north in the world with a population in 2000 just over 4500 people. It is well within the zone of continuous permafrost located at the point of a peninsula marking the northern reaches of the state. The Chukchi Sea lies to the west and the Beaufort seas to the east. The bluffs along the west coast of the peninsula are considerable higher than those on the east. The west coast bluffs are composed of interbedded sands and gravels mixed with ground ice, while the bluffs to the east along Elson Lagoon are considerably lower and are dominated by ice-rich, silts and sands. Manley (2004) determined erosion rates along the peninsula over a fifty year time period. It was determined that 91% of the coastline of the peninsula had undergone erosion over this time period and that the average rate of coastal erosion was 0.91 m/yr with an average horizontal displacement of 42.5 m. Lower rates of erosion were found along the higher bluffs on the Chukchi coast (0.3 m/yr) while much higher rates were found along the Elson Lagoon coast (up to 5 m/yr). Along the portion of coastline adjacent to the town of Barrow, the shoreline eroded 0.2 to 0.8 m/yr. Overall the peninsula decreased in area by 11.8 $km^2$. The rates of erosion in Barrow have had some impact on the structures in the community, however, there are other communities in this region experiencing far greater problems as a result of coastal erosion.

Point Lay a community located about 300 km southwest of Barrow was forced to relocate in the 1970s. Prior to this, the village had established itself on a barrier island just off the coast of the mainland. Barrier islands are extremely susceptible to storm surges and they have a natural tendency to migrate landward as the sediments are removed from the ocean side and redeposited on the mainland side.

Point Hope is another village that has had to face consequences of coastal erosion and relocation. Point Hope was located near the tip of a spit about 200 km southwest of Point Lay. This portion of the coastline experienced a great deal of erosion during the 20$^{th}$ Century and by the 1 960s nearly two-thirds of the village was lost (Mason et. al., 1997). In the mid-1970s the remainder of the town was relocated several miles eastward along the spit.

Kivalina, a town located on a barrier island southeast of Point Hope lies only 1 to 2 feet above normal high- tide line. The island is believed to be eroding on both sides, ocean and mainland, at estimated rates in excess of 1 to 1.3 meters per year (Mason et. al., 1997). A large storm struck the town in October of 2004 and the idea of relocation has risen to the forefront.

Shishmaref, another town located on a barrier island (Sarichef Island) about 275 km southwest of Kivalina, on the other side of the bay, is also experiencing extreme erosion problems. Over the last 10-20 years, locals believe that around 9 meters of erosion has occurred on the dune scarp in town and the island has lost nearly 35 meters of beach front in the past 40 years (Mason et. al., 1997). In a recent report by the Army Corps of Engineers there are four alternatives that the village of Shishmaref faces, either stay on Sarichef Island, relocate to a new mainland site, collocate to Nome, or collocate to Kotzebue. The cost of each of these alternatives would be $110 million, $180 million, $95 million, and $140 million, respectively.

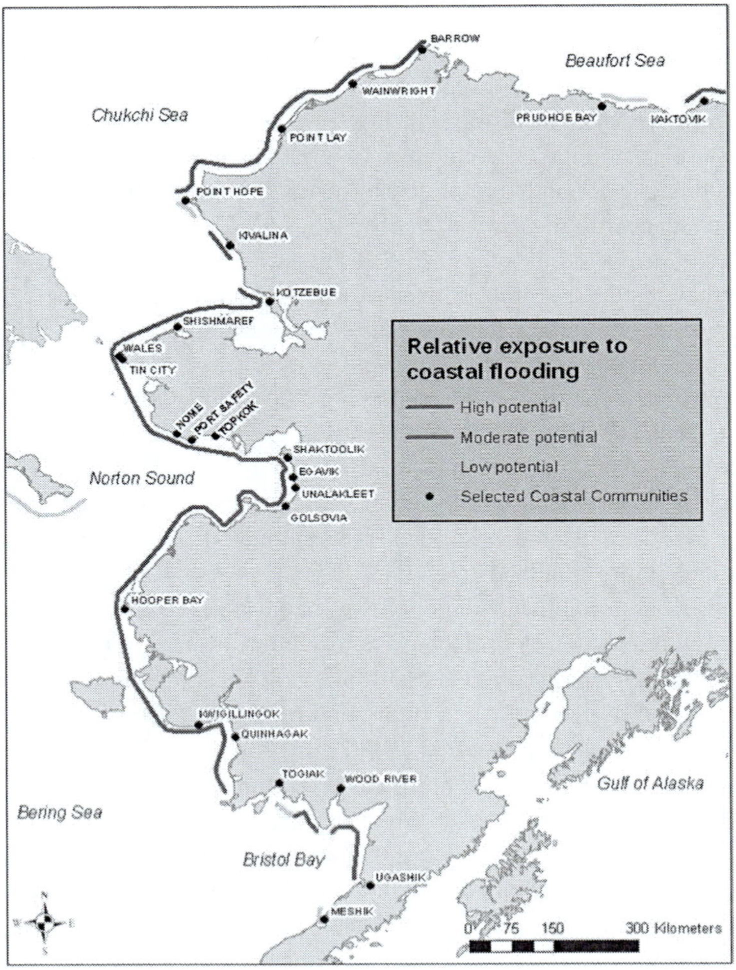


Figure 23. Map showing susceptibility to flooding during coastal storms. Low: flooding not common, Moderate: flooding every 3 to 5 years, sufficient to cause beach erosion, High: significant coastal erosion at least every 2 years (modified from Mason et al., 1997).

Following current climate conditions many of the islands and stretches of coastline in this region could experience higher rates of erosion resulting from an increased summer fetch due to more open ocean conditions. This combined with enhanced permafrost degradation could further the negative impacts occurring in these arctic villages. As these communities have become fixed settlements the costs associated with their relocation, both financially and culturally have drastically increased.

## Anthropogenic Drivers

The anthropogenic drivers contributing most directly to landscape changes are a result of resource exploration, urban development, and logging. Primary drivers in this context are discussed below.

### *Oil and Gas Exploration and Development*

The idea that Alaska was rich in oil and gas reserves has existed for quite sometime. The first wells drilled for oil extraction occurred at the end of the 1 800s on the lower Kenai Peninsula and the first commercial discovery of oil came in the mid-1950s at the Swanson River oil field on the northern region of the Kenai Peninsula. This was followed by discoveries on the Kenai Peninsula and in Cook Inlet. The Cook Inlet region provided the first major quantity of profitable oil extraction and a mini-boom in population occurred on the Kenai Peninsula in the late 1950s/early 1960s. The largest oil field in North America was discovered on the North Slope in 1968. To transport the reserves southward from the North Slope the Trans-Alaska Pipeline System (TAPS) was completed in 1977 and since that time the modern economy of the state has been closely tied to the oil and gas prices worldwide. With the North Slope crude oil production peaking in late 1980s (Error! Reference source not found.), there is a mounting interest to develop the natural gas reserves within Alaska in the near future.

### *Urban Development*

Urban centers and large towns are not typical of Alaskan communities. There is one sizeable population center in Alaska, the Municipality of Anchorage (MOA) which represents 43% (277,498 people) of the state's population and consists of Anchorage, Eagle River-Chugiak, and Girdwood (AK Community Database, 2005). Juneau, the state capital, represents the second largest population center in the state with a population around 31,000, while Fairbanks has the third largest population at 30,000. The only other town in Alaska with a population greater than 10,000 people is College which is essentially an integral part of Fairbanks. In contrast, there are 25 towns in Alaska with a population between 3,000 people and 10,000 people.

### *Selected Communities*

#### Anchorage

Much of the growth that has occurred in the Anchorage area began in the 1940s and has been driven by the boom and bust cycles that have dominated population trends statewide (Figure 10). For example, the population of Anchorage increased more than 40% during state oil revenue boom of the 1980s (Goldsmith et al., 2005).

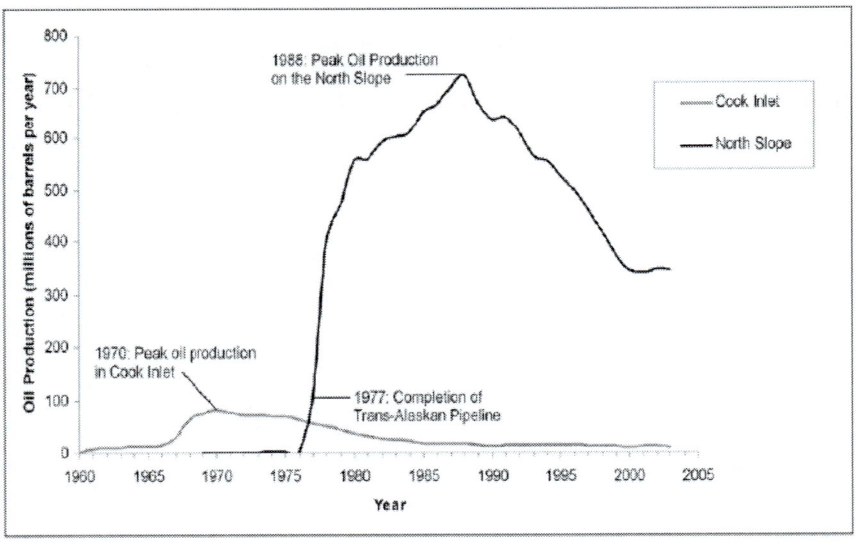


Figure 24. Graphs of oil production (millions of barrels per year) for the state. Cook Inlet production years range from 1960s to 2003. North Slope oil production ranges from 1969 to 2003 (data source: Alaska Department of Natural Resources).

A recent study conducted by Markon (2003) revealed that the Greater Anchorage Bowl, an area of approximately 300 $km^2$ consisted of approximately 24% urban area in the early 1970s. By the year 2000 the percentage of urban lands increased to 45%, almost a factor of 2. Much of the development had occurred in areas that proved to be relatively easy surfaces in which to construct upon (i.e. flat, relatively dry ground). For Anchorage, the majority of the urban growth took place between the late 1970s and the early 1980s, accounting for 15% of the 21% change over the thirty year period.

### Fairbanks

The population of Fairbanks has grown dramatically over the past century as well. In 1950, the population was slightly higher than 5,000 people and by 2000 the population was over 30,000 people, an increase of 600% (U.S. Census Bureau, 2000). To date, a detailed study has not been conducted on the change in urban areas within the Fairbanks region; however, there has been a study on the development of an urban heat island effect in Fairbanks over the time period 1949 to 1997 (Magee et al., 1999). The urban heat island effect describes the effects of an urban area (development and human activity) on temperatures and it can serve as a proxy of the urbanization of an area. In the study by Magee et al. (1999), the mean annual heat island effect observed at Fairbanks grew by 0.4°C from 1949 to 1997, with the winter experiencing an increase of 1.0°C.

### The Matanuska-Susitna Valley

The Mat-Su Valley has become an area of rapid urbanization as it is essentially an exurb of the Municipality of Anchorage. The Mat-Su area is also the primary agricultural region in the state. Thus, the valley has been experiencing the conversion of agricultural lands to residential and other urban areas since the early 1980s (Alaska Department of Commerce, 2005). In the 1960s the Mat-Su Borough population was around 5,000 people and by 2005 it has grown to over 74,000 people, an increase of nearly 1500% over the 45 year time period (Alaska Department of Commerce, 2005).

### *Mining*

Mining has recently become the second most valuable commodity in the state with the most important economic minerals mined in the state, in order of importance, are zinc, gold, lead, coal, sand and gravel, building stone, silver, copper, and jade. Since 1981, the total value of the mineral industry has increased from $290 million to $1.4 billion in 2004, with the industry exceeding $1 billion per year since 1996 (Figure 25). In 2004, 21 mines across the state accounted for 97% of the money spent on mineral exploration, and 59% of the money spent on exploration occurred in southwestern Alaska.

## Logging

There are two National Forests in Alaska; the Tongass and the Chugach. The Tongass National Forest is 67,200 km$^2$, of which 40,000 km$^2$ are forested and 14,500 km$^2$ are deemed harvestable and available to commercial activities with the target species being Sitka spruce, western hemlock, western red cedar, and yellow cedar. The Chugach National Forest spans 24,000 km$^2$. Of this area only 6% is considered to be harvestable forest and the target species are Sitka spruce and white spruce (USDA – Forest Service).

Commercial logging began in southeast Alaska during the 1 920s and it boomed in the Tongass National Forest during the 1950s where the industry dominated. Historically, the timber industry grew through the 1980s, butby the late 1990s, the reduced harvests led to closure of pulp mills which caused a fifty percent reduction of the work force (Leask et al., 2001). Figure 26 shows the trend in area harvested in the Tongass National Forest and the Chugach National Forest from 1994 to 2004.

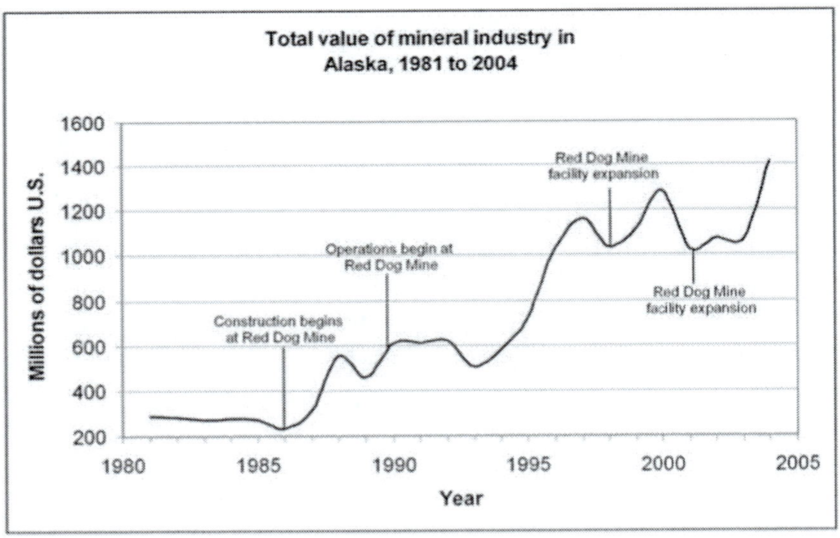


Figure 25. The total value of the mineral industry (non-petroleum) in Alaska from 1981 to 2004. The Red Dog Mine operation (largest zinc producer in the world) began in 1990 and much of the growth in this industry is presumed to be a result of this mine (data source: Division of Geological and Geophysical Surveys).

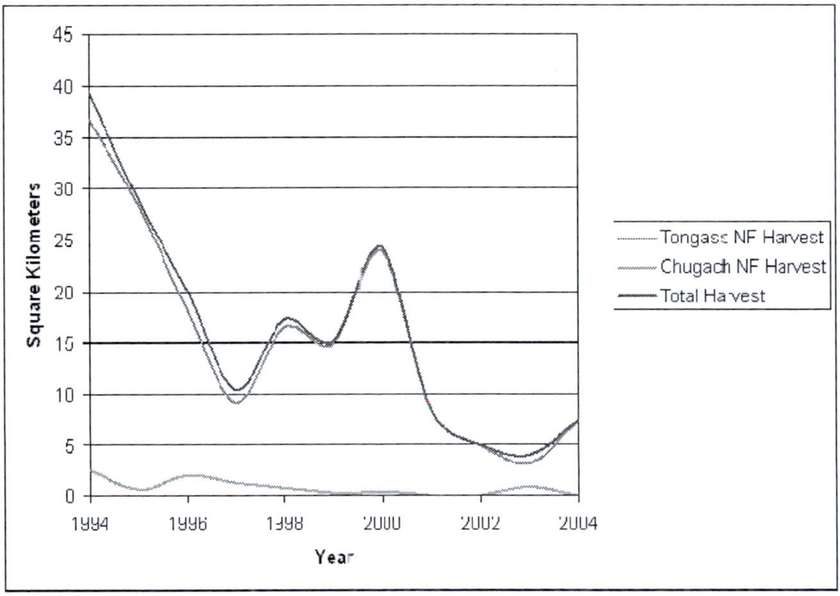


Figure 26. Area of forest harvested in the two national forests in Alaska from 1994 to 2004. The Tongass National Forest represents the majority of the logging taking place in the NF in Alaska (data source: US Forest Service).

Logging has also occurred on the Kenai Peninsula as a response to the spruce bark beetle outbreak that ravaged the forests in the mid-1990s. The spruce bark beetle infestation on the Kenai Peninsula is considered to be the most devastating insect infestation to occur in North America. Since 1989 the U.S. Forest Service and the Alaska Division of Forestry have mapped over 5,600 km$^2$ of beetle infected forests on the Kenai Peninsula. The infestation of the 1990s accelerated in 1992 and peaked in 1996, however, new stands still continue to be infested each year. The Kenai Peninsula Borough (KPB), Spruce Bark Beetle Mitigation Program (SBBMP) identified 133,000 acres of harvested forest on the Kenai by using satellite imagery from 2000 to 2005 (Figure 27). The boreal forests of the interior have been logged to a much lesser degree, but there is increased interest in harvesting the resources there. The target species in the boreal forest are white spruce, quaking aspen, and paper birch.

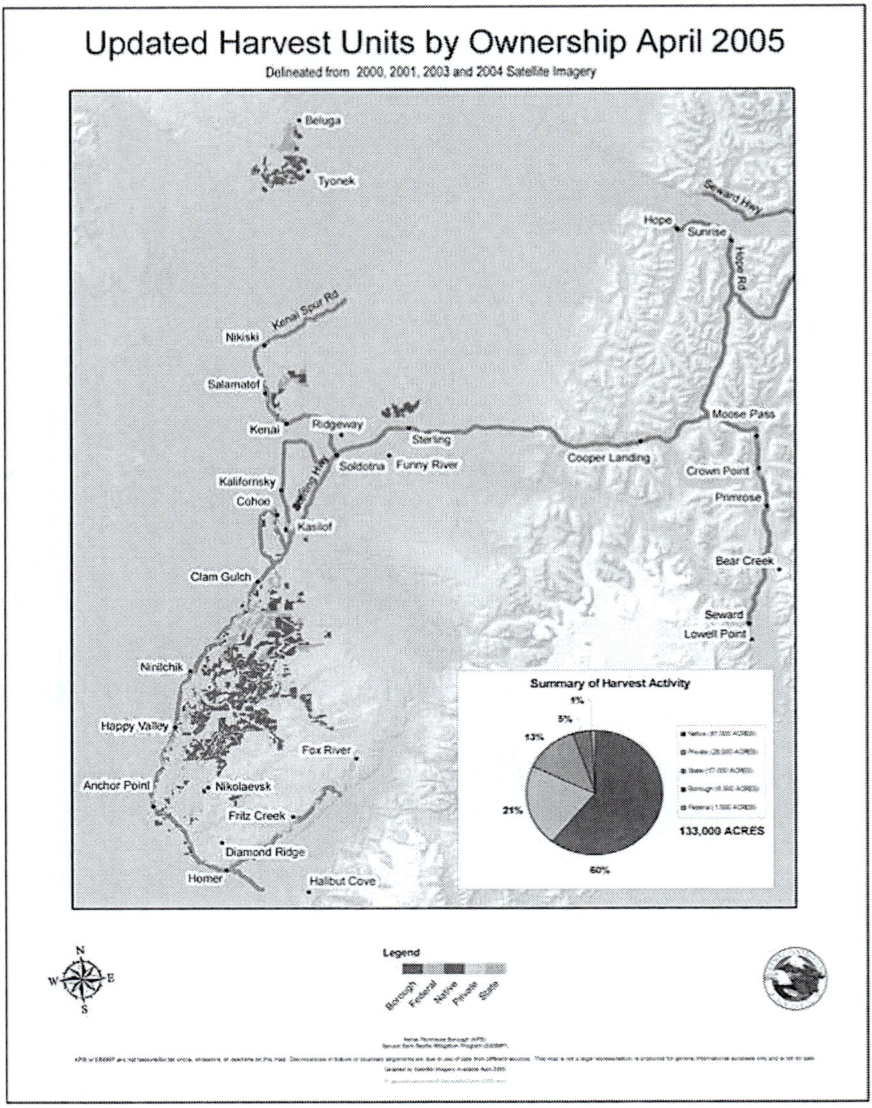


Figure 27. Harvested forest land on the Kenai Peninsula as a response to the spruce bark beetle activity, as of April 2005. The map does not display exact years of forest harvest, and thus, some of the harvested forest could have occurred pre-2000 (Source: Kenai Peninsula Borough).

# Discussion and Conclusions

The land surface of Alaska is sparsely populated and the impacts from humans are far less extensive when compared to the contiguous United States. The changes in the population and the economy of Alaska have historically been driven by boom and bust cycles, primarily from mineral discoveries, logging, military expansion, and oil and gas development; however, the changes as a result of these factors have occurred in relatively small, localized areas. The economy of Alaska changed drastically after the discovery of the North Slope oil fields. This was followed by an increase in population partially because of TAPS construction, but also because of an increase in state spending of oil revenue in the 1980s (Goldsmith et al., 2005; Markon, 2003). The majority of the state's revenues still come from this resource; however, at the present time, much of the growth occurring in the state is a result of tourism and a more stable population, one which exhibits a longer residence time (Leask et al., 2000). Many of the statewide changes taking place in the land surface however, are a result of natural or climate driven processes as opposed to direct anthropogenic activities.

The widespread statewide changes appear to be occurring in relation to changing climatic conditions. The mean annual air temperature has warmed by about 2.0°C since the middle of the 20th Century. Some parts of the state have experienced more warming than others but the general trend is apparent statewide. Many areas have also experienced more mean annual precipitation over that time period with the exception of the Arctic. The weather related phenomena over the past ~ 50 years are believed to be the cause of many of the changes taking place across the state today, such as more intense fire seasons, lightning and thunderstorms, insect infestation outbreaks, glacier retreat, permafrost degradation, and enhanced coastal erosion (Juday 1998; Osterkamp and Romanovsky, 1999).

During 2004 and 2005 Alaska has experienced the first and third largest fire seasons on record and these have been attributed to summer drought conditions and an increase in lightning activity (McGuiney et al., 2005; Richmond and Shy, 2005). The majority of the fires have occurred in the interior of the state. However, if the current warming scenario continues the expansion of forest fires to the coastal forest zones could become more prevalent.

The impact of insect infestation outbreaks is another component of land surface change occurring in Alaska in recent times. The recent spruce bark beetle attack (late 1980s to 1998) has had the largest impact on the integrity of

Alaska's forests on the Kenai Peninsula and the Copper River region. The beetles leave behind a forest filled with dead spruce timber, potentially creating an instant fuel source for forest fires. As a consequence logging has become more proactive on the Kenai since the peak in the bark beetle outbreak of the mid-1990s as a means of timber salvage and hazard remediation. In 2004, spruce bark beetle infestations were reported on the Seward Peninsula, east of Nome, in western Alaska and if current climate trends continue the beetles range could expand northward. (USFS-Forest Health Monitoring, 2005).

Glacier retreat and permafrost degradation are expected to continue in the near future. The land terminating glaciers appear to be more closely linked to changing climate conditions than the tidewater glaciers and they serve as a proxy of how rapid changes can occur. For example, two of the most visited tourist attractions in the state, Portage Glacier and Exit Glacier, have been in a continual state of retreat in recent times. Permafrost degradation is expected to have the biggest impact in the interior region of the state were it is more susceptible to changing climatic conditions because the permafrost is less extensive and thinner than in continuous zones, which should result in basal thawing (Osterkamp and Romanovsky, 1999).

Coastal erosion is expected to increase along Alaska's western and northern coastlines following current climate trends. These coastlines consist primarily of unconsolidated sediments in permafrost, thus, they are susceptible to erosion. An increase in fetch and storm activity is expected because the pack ice extent is decreasing in the Bering Sea and Arctic Ocean. In recent times, the pack ice has resided farther from land and for a longer part of the year.

Coastal erosion along these coastlines has received a lot of press in recent times due to effect on local communities. The people of this region used to live a lifestyle in which mobility and relocation allowed them to cope with the changing conditions of the coastline. Today, however, modern amenities such as plumbing, fixed structures, and power plants have decreased their mobility significantly. Several communities along these coastlines have already had to deal with the loss of land due to coastal erosion and several more will probably face this consequence in the future.

The anthropogenic changes that have occurred in Alaska have remained localized in nature and relatively small in scale. The largest anthropogenic impacts on land cover change have come from resource exploration/extraction and urbanization. The development of the North Slope oil reserves has had the largest impact on the state's economy and in the near future an adjacent gas pipeline is planned for development along side the Trans-Alaskan Pipeline, thus harboring the potential for another boom in Alaska. The mining (non-petroleum) industry has recently become the second most valuable commodity in the state behind oil and gas and the potential for future development is likely as exploration is underway in areas of the Brooks Range as well as areas in Southwestern Alaska. The utilization of forest resources in Alaska was highest during the logging boom of the 1950s in southeastern Alaska however, this boom began to wane in the late 1 980s. Urbanization will probably continue to have an effect on land-cover change as the population of Alaska is expected to grow throughout the 21st Century (Campbell, 1996). The Municipality of Anchorage is the largest urban area in the state, however much of the developable lands in Anchorage have already been developed, as a result new growth in the Cook Inlet Region is expected to continue to occur in the Matanuska-Susitna Valley.

It is apparent that several changes are occurring in the land surface of Alaska. However, all of the information and data available covers a relatively short time period, roughly the 1950s onward. It will be important to extend the monitoring of these changes into the future to compare rates of change from the latter half of the 20th Century, into the 21st Century.

## LITERATURE CITED

[1] Alaska Climate Research Center. (2005). http://climate.gi.alaska.edu/. Atmospheric Sciences Group, Geophysical Institute, University of Alaska Faribanks.

[2] Alaska Department of Commerce. (2005). Alaska Communities Database Online. http://www.commerce.state.ak.us/.

[3] Alaska Department of Natural Resources. (2002). Fact Sheet: Land Ownership in Alaska. http://www.dnr.state.ak.us/mlw/factsht/land_own.pdf.

[4] Alaska Earthquake Information Center. (2005). http://www.aeic.alaska.edu/.Geophysical Institute ,University of Alaska Fairbanks.

[5] Alaska Fire Service. (2004). fire.ak.blm.gov/unique/ docs/ffr/AICC%2004%20overview.asp.

[6] Arendt, A. A., Echelmeyer, K. A., Harrison, W. D., Lingle, C. S. & Valentine, V. B. (2002). Rapid wastage of Alaska glaciers and their contribution to rising sea level. *Science.*, *297*, 382-386.

[7] Bernd-Cohen, T. & Gordon, M. (1999). State coastal program effectiveness in protecting natural beaches, dunes, bluffs and rock shores. *Coastal Management*, *27*, 187-217.

[8] Blanchet, D. (2002). Interview, Reuters News Service. *Experts confirm observation of glaciers' fast melt*, story by Yereth Rosen.

[9] Brown, C. S., Meier, M. F. & Post, A. (1982). Calving speed of Alaska tidewater glaciers, with application to Columbia Glacier: U.S. Geological Survey Professional Paper 1258-C, C1-C13, 4pls.

[10] Campbell, P. R. (1996). Population Projections for States by Age, Sex, Race, and Hispanic Origin: 1995 to 2005. http://www.census.gov/population/www/projections/ppl47.html#trends.

[11] Cameron, C. E. (2005). Latitude and Longitudes of volcanoes in Alaska. State of Alaska, Department of Natural Resources, Division of Geological and Geophysical Surveys, tabular digital data. http://www.avo.alaska.edu/volcanoes/latlong.php.

[12] Court, A. & Griffiths, J. F. (1992). Thunderstorm Climatology. In *Thunderstorm Morphology and Dynamics, vol. 2*, E. Kessler, Ed., *University of Oklahoma*, 9-40.

[13] Curtis, J., Wendler, G., Stone, R. & Dutton, E. (1998). Precipitation decrease in the Western Arctic, with special emphasis on Barrow and Barter Island, Alaska. *International Journal of Climatology.*, *18*, 1687-1707.

[14] Cusick, J. (2001). Foliar nutrients in black cottonwood and Sitka alder along a soil chronosequence at Exit Glacier, Kenai Fjords National Park, Alaska. *Masters thesis*, University of Alaska, Anchorage.

[15] Derksen, D. (2005). Avian population response response to ecological change along the arctic coastal plain. *Alaska Science Center, U.S. Geological Survey Progress Report,* permafrost temperature figure: Gary Clow.

[16] Dissing, D. & Verbyla, D. L. (2003). Spatial patterns of lightning strikes in interior Alaska and their relations to elevation and vegetation. *Canadian Journal of Forestry Research., 33*, 770-782.

[17] Dyurgerov, M. B. & Meier, M. F. (2000). Twentieth century climate change; evidence from small glaciers. *Proceedings of the National Academy of Sciences of the United States of America*, *97(4*), 1406–1411.

[18] Erskine, J. ed. (2005). Alaska Economic Trends: The Year 2004 in Review. *Alaska Department of Workforce Development*, *25(4)*, 23.

[19] Global Insight. (2004). The Alaska Tourism Satellite Account: A comprehensive analysis of the economic contribution of travel &

tourism, prepared for the Alaska Department of Commerce, Community and Economic Development, p. 24. http://www.dced.state.ak.us/oed/toubus/.

[20] Goldsmith, S., Howe, L. & Leask, L. (2005). Anchorage at 90: Changing fast, with more to come. *Institute of Social and Economic Research, University of Alaska Anchorage*, Research Summary No. 4.

[21] Goldsmith, S., Angvik, J., Howe, L., Hill, A. & Leask, L. (2004). The Status of Alaska Natives Report. *Institute of Social and Economic Research, University of Alaska Anchorage*, v.*1*.

[22] Hall, D. K., Benson, C. S. & Field, W. O. (1995). Changes of glaciers in Glacier Bay, Alaska, using ground and satellite measurements. *Physical Geography*, *16(1)*, 27-41.

[23] Hall, D. K. (2005). Monitoring glacier changes from space. http://www.isset.org/site_of_the_month/glacier_bay/hall.science.txt.html

[24] Hartmann, B., and G. Wendler. The significance of the 1976 Pacific climate shift in the climatology of Alaska. *Journal of Climate*, *18(22)*, 4824-4839.

[25] Jorgenson, M. T., Racine, C. H., Walters, J. C. & Osterkamp, T. E. (2001). Permafrost degradation and ecological changes associated with a warming climate in central Alaska. *Climatic Change*, *48(4)*, 551-579.

[26] Juday, G., Ott, R., Valentine, D. & Barber, V. (1998). Forests, climate stress, insects, and fire. In, Implications of Global Change in Alaska and Bering Sea Region. *Proceedings of a workshop, University of Alaska Fairbanks, June 1997*. G. Weller, & P. Anderson, eds. Center for Global Change and Arctic System Research, University of Alaska Fairbanks, 23-49.

[27] Leask, L., Killorin, M. & Martin, S. (2001). Trends in Alaska's people and economy. *Institute of Social and Economic Research, University of Alaska Anchorage*, 16.

[28] Lachenbruch, A. H. & Marshall, B. V. (1986). Changing climate: Geothermal evidence from permafrost in the Alaskan arctic. *Science*, *234*, 689-696.

[29] Magee, N., Curtis, J. & Wendler, G. (1999). The urban heat island effect at Fairbanks, Alaska. *Theoretical and Applied Climatology*, *64*, 39-47.

[30] Mantua, N. J. (2002). The Pacific Decadal Oscillation *in The Encyclopedia of Global Environmental Change, Vol. 1, The Earth System: physical and chemical dimension of global environmental change*, 592-594. Wiley & Sons, Ltd: Chichester.

[31] Mason, O., William, J. & Pilkey, O. H. (1997). *Living with the Coast of*

*Alaska*. Duke University Press: Durham, NC.
[32] McGuiney, E., Shulski, M. & Wendler, G. (2005). Alaska lightning climatology and application to wildfire science. *Conference on Meteorological Applications of Lightning Data.* San Diego, CA.
[33] Meier, M. F. (1984). Contribution of small glaciers to global sea level. *Science, 226(4681)*, 1418-1421.
[34] Meier, M. F. & Post, A. (1987). Fast tidewater glaciers. *Journal of Geophysical Research, 92(B9)*, 9051-9058.
[35] Miller, T.P., McGimsey, R. G., Richter, D. H., Riehle, J. R., Nye, C. J., Yount, M. E. & Dumoulin, J. A. (1998). Catalog of the Historically Active Volcanoes of Alaska. *U.S. Geological Survey Open-File Report 98-582*, 104.
[36] Muller, S. W. (1947). *Permafrost or Permanently Frozen Ground and Related Engineering Problems.* J.W. Edwards, Ann Arbor, Michigan, *136*.
[37] Northern Economics Inc. (2004). Alaska Visitor Arrivals: Fall/Winter 2003-2004, prepared for the State of Alaska, Department of Commerce, Community and Economic Development, *31,* http://www.commerce.state.ak.us/oed/toubus/research.htm.
[38] Northern Economics Inc. (2004). Alaska Visitor Arrivals: Summer 2004, prepared for the State of Alaska, Department of Commerce, Community and Economic Development, *32*, http://www.commerce.state.ak.us/oed/toubus/research.htm.
[39] Osterkamp, T. E. & Romanovsky, V. E. (1999). Evidence of warming and thawing of discontinuous permafrost in Alaska. *Permafrost and Periglacial Processes, 10(1)*, 17-37.
[40] Post, A. & Viens, R. J. (1994). *Preliminary bathymetry of Shoup Basin and late Holocene changes of Shoup Glacier*, Alaska: U.S. Geological Survey Water-Resources Investigation Report, 94-4093, 11.
[41] Richmond, M. & Shy, T. L. (2005). An extraordinary summer in the interior of Alaska. *Eighth Conference on Polar Meteorology and Oceanography* (San Diego, California), American Meteorological Society.
[42] Schmidt, R. A. M. (1961). Recession of Portage Glacier, Alaska. U.S. Geological *Survey Professional Paper*, *358*, P0424-D, D202-D203.
[43] Seramur, K. C., Powell, R. D. & Carlson, P. R. (1996). Evaluation of conditions along the grounding line of temperate marine glaciers: an example from Muir Inlet, Glacier Bay, Alaska. *Marine Geology, 140*, 307-327.

[44] Serreze, M. C., Walsh, J. E., Chapin III, F. S., Osterkamp, T., Dyurgerov, M., Romanovsky, V., Oechel, W. C., Morison, J., Zhang, T. & Barry, R. G. (2000). Observational evidence of recent change in the northern high-latitude environment. *Climatic Change*, *46(1-2)*, 159-207.

[45] Shulksi, M. D., Wendler, G. & Hartmann, B. M. (2005). Climate trends and variability in Alasks since 1950. *Proceedings of the 15th Conference on Applied Climatology*. Savannah, GA.

[46] State of the Beach. (2005). http://www.surfrider.org/stateofthebeach/.

[47] Stover, C. W. & Coffman, J. L. (1993). Seismicity of the United States, 1569-1989 (Revised). *USGS Professional Paper*, *1527*, 418.

[48] Sturm, M., Hall, D. K., Benson, C. S. & Field, W. O. (1991). Non-climatic control of glacier-terminus fluctuations in the Wrangell and Chugach Mountains, Alaska, U.S.A. *Journal of Glaciology*, *37(27)*, 348-356.

[49] Trabant, D. C., March, R. S. & Thomas, D. S. (2003). Hubbard Glacier, Alaska: Growing and advancing in spite of global climate change and the 1986 and 2002 Russell Lake outburst floods. *U.S. Geological Survey Fact Sheet* 001-03.

[50] U.S. Census Bureau. (2005). http://www.census.gov/.

[51] USFS. (1998). *Forest Health Protection Repor, Forest Insect and Disease Conditions in Alaska,* Anchorage, Alaska: Alaska Department of Natural Resources, cd-rom.

[52] USFS-Forest Health Conditions in Alaska, (2004). *Forest Health Conditions in Alaska – 2004: A Forest Health Protection Report.* USDA, USFS, and Alaska DNR. 97.

[53] van Everdingen, R., ed. (2005). "Multi-language glossary of permafrost and related ground-ice terms." from http://nsidc.org/fgdc/glossary/.

[54] Werner, R. A. (1996). Forest health in boreal ecosystems of Alaska. *The Forestry Chronicle.*, *72(1)*, 43-45.

[55] Wesser, S., Allen, J., Murray, M., Markon, C., Fleming, M., Binnian, E., Brunt, K., Winterberger, K., Kelley, T. & Burgress, W. (2001). Stand and landscape level analyses of a spruce bark beetle (Dendroctonus rufipennis (Kirby)) infestation within Wrangell-St. Elias National Park and Preserve. *WRST Technical Report, 99-01, cd-rom.*

In: Faces of Alaska
Editor: Mark C. Monteith
ISBN: 978-1-61122-733-8


*Chapter 2*

# ALASKA'S RENEWABLE ENERGY POTENTIAL

***Sandia National Laboratories***

## ABSTRACT

This paper delivers a brief survey of renewable energy technologies applicable to Alaska's climate, latitude, geography, and geology. We first identify Alaska's natural renewable energy resources and which renewable energy technologies would be most productive. We survey the current state of renewable energy technologies and research efforts within the U.S. and, where appropriate, internationally. We also present information on the current state of Alaska's renewable energy assets, incentives, and commercial enterprises. Finally, we describe places where research efforts at Sandia National Laboratories could assist the state of Alaska with its renewable energy technology investment efforts.

## ACKNOWLEDGMENTS

This report was compiled from many sources. Where a specific report was directly quoted, the citation has been provided. However, the body of this report summarizes and describes research conducted worldwide. The description contained herein would not be possible absent the vast research efforts worldwide *and* the efforts of those people and organizations in making their results available to all.

Many renewable-energy (RE) resources can be grouped into five categories, based on the technology used to extract, capture, or exploit them. The categories are solar, wind, geothermal, biomass, and hydropower. For Alaska, the most applicable RE technologies are wind, geothermal, and hydropower (both inland/river hydropower and ocean/tidal power).

## ACRONYMS

| | | | |
|---|---|---|---|
| AC | alternating current | NBB | National Biodiesel Board |
| AEA | Alaska Energy Authority | NRBP | Northeast Regional Biomass Program |
| BEOpt | Building Energy Optimization | NREL | National Renewable Energy Lab |
| BES | (DOE Office of) Basic Energy Sciences | NWCC | National Wind Coordinating Committee |
| BOS | balance of system | NWTC | National Wind Technology Center |
| CETC | CANMET Energy Technology Centre | OEDER | (DOE) Office of Electricity Delivery and Energy Reliability |
| CRADA | cooperative research and development agreement | PDC | polycrystalline diamond compact |
| DC | direct current | PSEL | Photovoltaic Systems Evaluation Lab |
| DNR | Department of Natural Resources | R&D | research and development |
| DOE | U.S. Department of Energy | RE | renewable energy |
| DWD | diagnostics-while-drilling | REAP | Renewable Energy Alaska Project |
| EER | energy-efficiency ratio | RFA | Renewable Fuels Association |
| EERE | (DOE Office of) Energy Efficiency and Renewable Energy | SNAP | sustainable natural lternative power |
| EES | Electrical Energy Storage (research prog.) | STAR | sweep-twist adaptive rotor |
| GHG | greenhouse gas(es) | TRNSYS | Transient System Simulation |
| GVEA | Golden Valley Electric Association | UAF | University of Alaska–Fairbanks |
| IEA | International Energy Agency | USDA | U.S. Department of Agriculture |
| INE | Institute of Northern Engineering | USFS | U.S. Forest Service |
| LBNL | Lawrence Berkeley | VHP | Virtual Hydropower Prospector |
| MOU | National Laboratory memorandum of understanding | WPA | Wind Powering America |

We have also identified leading institutions where RE research is underway today around the world and their general areas of specialty. This list could be used as a basis for developing a workshop on this topic and inviting some of the world's foremost researchers to Alaska.

## WHO ARE THE RESEARCH LEADERS?

A large body of international research exists on RE and energy conservation in cold-climate areas. Because the German government significantly subsidizes RE and preferentially solar power (~40¢/kWh for 20 years), a great deal of funding is available for research there. Canada also has a well-developed research base in cold-climate RE and energy conservation. In the U.S., Sandia National Laboratories (Sandia) and the National Renewable Energy Laboratory (NREL) in Golden, Colorado, are recognized as national leaders in RE research.

## WHAT ARE THE APPLICABLE TECHNOLOGIES?

The RE resource maps for Alaska at the U.S. Department of Energy (DOE) Energy Efficiency and Renewable Energy (EERE) program's Web site[†] do not show a large potential for solar power. Most of Alaska receives an annual average of 2000–3000 Whr/m$^2$ per day. Additionally, the long periods of dark or near dark further detract from this limited resource's annual reliability. The two methods for collecting solar *thermal* energy, light concentrators and using closed- loop water pipes, are not viable; the sun's energy at the Earth's surface is too weak to allow concentrators to be effective and the ambient temperature is too often too low to prevent the water pipes from freezing. Alaskans *can* exploit flat-panel photovoltaic solar collectors, but, given these systems' current costs and power-delivery densities, a stringent cost-benefit analysis would be necessary to determine their economic feasibility in any specific application.

What Alaska *does* possess is generous RE in the form of geothermal, wind, and hydropower resources that range from good to superb. Although Alaska's northern latitude and short growing season makes inland biomass

[†] http://apps1.eere.energy.gov/states/alternatives/resources

sources scarce, much of Alaska along the coast is beginning to look at biomass for co-generation facilities.

The RE resources just identified are rich enough to be exploited not only on the small scale (i.e., by domestic dwellings or small buildings) but for large-scale electrical power generation. Because exploiting RE resources at both of these scales involves very different technologies, the next section of this paper will describe the domestic-dwelling size applications. Then, we will present information on the large-scale power generation technologies.

## DOMESTIC/SMALL-SCALE RE POWER GENERATION AND CONSUMPTION

### Power Generation

The two largest energy usages in a typical home are electricity and heat (hot water and space heating). RE sources can address both of these energy paths. A study published in 2006[1] investigated five domestic-size RE configurations for cold climate regions (research was conducted at Hokkaido, Japan). They found that for regions with a climate similar to Alaska's, a system that uses photovoltaic panels for electrical generation and a heating/hot-water system that has

- a well-insulated electric water storage tank that services hot-water loads and
- a compact boiler/geothermal heat pump tank for room heating during the cold season

led to an ~28% reduction in greenhouse gas emissions. This system nearly completely eliminated the need for fuel oil.

Wind power is available to the typical Alaskan homeowner. Micro wind turbine technology is improving and more affordable. New microprocessor-based technology has increased performance, improved battery charging capability and unit reliability, and reduced "flutter" noise from the turbine blades. Statistics show the complementary nature of wind power and photovoltaics— wind speed often increases on cloudy days and after sunset.

**Table 1. Information on Typical Domestic RE Technology Applicable to Alaska**

| Typical RE Technology | A Supplier | General Performance Characteristics | Typical Industry Costs |
|---|---|---|---|
| Photovoltaic panels | SunPower Corporation (San Jose, CA) | From 9%–14% of available solar energy is delivered as electrical power. | ~$8–$1 1/W, installed |
| Geothermal heat pump | SunTeq Geo Distribu tors (central PA) | Heating: 3.3 EER†, cooling: 14.1 EER | ~$2,500 per ton of capacity (3 t/$7,500 for a typical home) |
| Micro wind turbine | Southwest Wind-power (Flagstaff, AZ) | A turbine with a rated capacity of 400 W and a startup wind speed of 8 mph can produce an estimated 38 kWh per month at 12 mph. | ~$600 for a unit with a rotor diameter of ~4'. |
| Micro hydro free-flow turbine | New Energy Corpora- tion, Inc. (Calgary, Alberta, Canada) | A vertical-axis turbine with a rated capacity of 5 kW/h when immersed in a stream flow of 3 m/s. | ~$28,000 for a turbine with 4 0.75 m rotor blades on a 1.5 m diam. |

Because of Alaska's large natural variation in elevation and abundance of river courses, micro hydropower is also feasible. Micro hydro installations are generally either free-flow generation where a turbine is immersed in the river flow or an engineered system where the water is diverted from the stream (weir) to flow through a pipe (penstock) to a turbine and then returned to the stream (tailrace). Free-flow turbines are less efficient, but more affordable because of the reduced infrastructure costs. Neither application would require a dam or reservoir. Implementing micro hydropower requires detailed site-specific measurements and plans.

It should be noted that photovoltaic, wind-turbine, and micro-hydro technologies require significant support equipment (referred to as the balance of system [BOS]: power inverters, battery storage, power distribution [transmission lines], etc.). These BOS costs must be considered during evaluation to decide whether to purchase/install an RE power system.

† EER (energy-efficiency ratio): the ratio of the system's total heating/cooling capacity to electrical energy input.

## Energy Conservation

Fossil-fuel-based energy systems often deliver higher specific energy and power content than RE technologies due to the fuel's intrinsic storage. Therefore, of equal importance with installing and using renewable, non-$CO_2$-emitting energy generation is energy conservation and the reduction of energy loss (increased efficiency).

Several modeling and simulation packages exist to help optimize a home's design to affordably reduce energy usage and increase conservation.[2] Included in these are the

- Transient System Simulation (TRNSYS) from the Solar Energy Laboratory at the University of Wisconsin,
- DOE2.2 Building Simulation Software developed by Lawrence Berkeley National Lab, and
- Building Energy Optimization (BEOpt) program developed by NREL.

These packages recommend constructing a superinsulated building envelope with two key features: a double-wall construction with a gap of at least 4" between the inner and outer wall[†] and roof trusses made with a "raised heel" that accommodates 2' of blown-in insulation. This truss design also permits longer roof overhangs—reducing passive solar heating in the summer.

These homes also used a solar water heating system (with a back-up tankless gas-fired system). In these case studies, the solar-heated water met over 60% of hot-water demand during the summer months.[3]

Alaska is home to the Cold Climate Housing Research Center (CCHRC)[‡] an industry-based, non-profit corporation dedicated to research that improves the durability, health, and affordability of shelter for people living in circumpolar regions around the globe. The CCHRC was created to facilitate the development, use, and testing of energy-efficient, durable, healthy, and cost-effective building technologies for Alaska and the world's cold-climate regions.

Alaska offers an excellent testing ground for cold-climate technologies and products. Its geography provides the full range of climatic conditions a researcher would encounter across the northern U.S.—from the windy, cool,

---

† This gap prevents the direct transfer of heat through the solid building members (wall studs).

‡ http://www.cchrc.org/

wet weather in the northeastern and northwestern states to the very cold, snowy conditions across the northern plains and Rocky Mountain regions. In addition, Alaska's cold season lasts for six months or longer in any given year, allowing ample time for researchers to conduct experiments and evaluate housing performance.

In September 2006, the CCHRC opened a new research and testing facility, located on the campus of America's Arctic University at the on the University of Alaska–Fairbanks (UAF), providing space to expand research as well as work more closely with university students, faculty, and researchers. The research center was conceived and developed by members of the Alaska State Home Builders Association, representing more than 1,200 building industry firms and groups.

The CCHRC has conducted extensive research on energy conserving building materials and construction methods. CCHRC research is presented on their Web site and is too detailed and extensive to be repeated here. This paper discusses CCHRC in further detail in the section dedicated to Alaska's existing RE assets (please see page 38).

# Large-Scale Renewable Energy Power Generation Potential in Alaska

## Geothermal Energy

Alaska has high-temperature geothermal resources that are suitable for electricity generation, as well as direct-use and heat-pump applications. The geologic and tectonic history of Alaska have produced substantial geothermal resources throughout the state. Recognized geothermal resources are concentrated in three regions:

1. within the "Ring of Fire," the volcanic arc that circles the Pacific and includes the Aleutian Islands, Alaska Peninsula, and Baranof Island;
2. a band of hot springs in central Alaska extending from the Seward Peninsula east to the Canadian border; and
3. near the Wrangell Mountains.

In 2003 the DOE, Alaska Energy Authority (AEA), and its contractors completed an assessment of geothermal resources in the state. The assessment identified substantial resources near Chena Hot Springs, Unalaska, and Akutan. In 2006 Chena Hot Springs installed a 400 kW geothermal power plant with the assistance of AEA's Denali Commission-funded Energy Cost Reduction program. The project eliminates the need for over 115,000 gallons of diesel fuel per year.

Despite Alaska's significant geothermal potential, the attributes of Alaska's geothermal resources remain poorly defined. AEA is involved in several task forces to better understand and develop these resources. These include

1. coordinating a statewide geothermal working group of industry, academic, and government officials interested in geothermal development;
2. participating in the DOE's Geopowering the West† program, which provides technical and financial support for western U.S. states; and
3. collaborating with the Alaska Department of Natural Resources (DNR) Division of Geological and Geophysical Surveys to compile geothermal data throughout Alaska.

## Wind Energy

DOE EERE program's Web site shows Alaska has sufficient wind resources for large- and small- scale wind power. Much of this resource is available near the state's large population/industrial centers—making transmission from point of generation to consumer relatively inexpensive.

## Hydroelectric Energy

Alaska has a very good hydroelectric power resource as a percentage of the state's electricity generation. For additional resource information, please review Idaho National Laboratory's Virtual Hydropower Prospector† (VHP). VHP is a convenient geographic information system tool designed to assist in locating and assessing natural stream water energy resources in the U.S.

† http://www1.eere.energy.gov/geothermal/gpw/

† http://hydropower.inel.gov/prospector/index.shtml

In the 2004 report "Water Energy Resources of the United States"[5] the DOE EERE estimates that the power potential of the U.S. water energy resources is ~300,000 MW—corresponding to an annual energy production of 2,680,000 GWh. Of this potential, ~40,000 MW, corresponding to the ~80,000 MW capacity of existing hydroelectric plants, have been developed. Power potential in zones that exclude new hydropower development accounts for about 90,000 MW. This leaves ~170,000 MW of potential or ~60% of the total that *has not been developed* and is not excluded from development. This potential power corresponds to an annual energy production of 1,501,500 GWh.

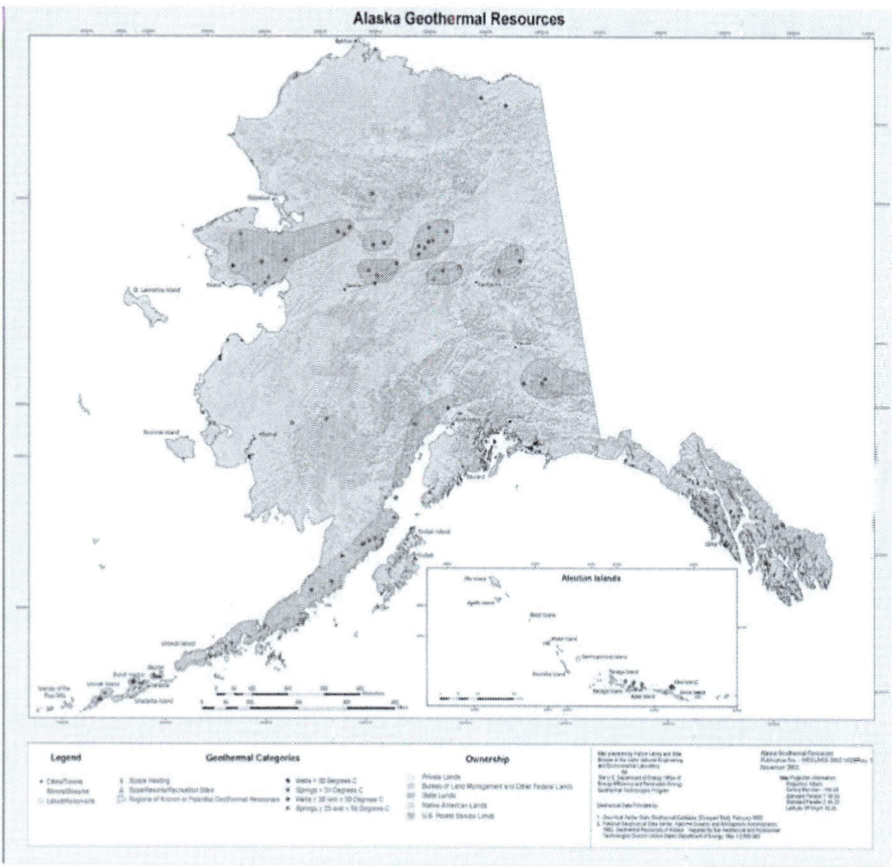


Figure 1. A geothermal-energy resource map for Alaska (source: AEA).

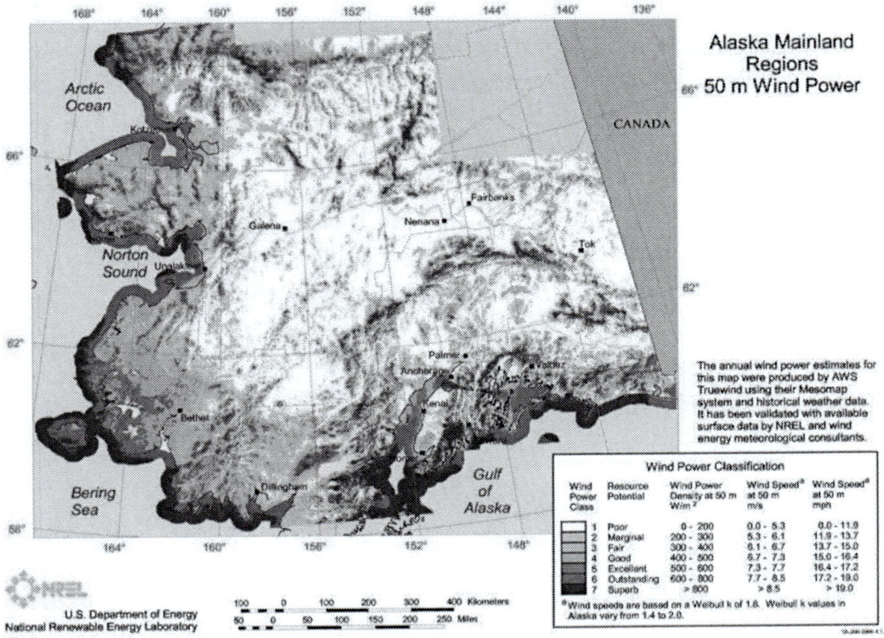


Figure 2. A wind-energy resource map for Alaska (source: NREL).

The study shows that *over half of the power potential of the U.S.* resides in Alaska (29%) and the Pacific Northwest (26%); in particular, in the states of Alaska, Washington, Idaho, and Oregon. Nearly half of the available (untapped) power potential also resides in Alaska (26%) and the Pacific Northwest (23%). Approximately 90% of this available potential is composed of high power potential (≥1 MW), high head/low power (head ≥30 ft and <1 MW) potential, and part of the low head/low power (head <30 ft and <1 MW) potential that could be realized using conventional turbine technology. However, the conventional turbine technology would have to be incorporated into new system configurations and not require impoundments to be determined by future research and development (R&D).

The boundary between the high power and low power classes defined by hydraulic head and flow rate is shown graphically in Figure 5(a) below. The low head/low power class is defined by the following two criteria

- all power potential <100 kW (microhydro) and
- power potential ≥100 kW but <1 MW with hydraulic head <30 ft.

The low head/low power class shown in Figure 5(a) is divided into the operating envelopes of three classes of low head/low power technologies

- conventional turbines—power ≥100 kW, but <1 MW *and* hydraulic head <30 ft, but ≥8 ft
- unconventional systems—power ≥100 kW, but <1 MW *and* hydraulic head <8 ft.
- microhydro technologies—power <100 kW

These operating envelopes are shown graphically in Figure 5(b).

The estimated, available, low head/low power potential of ~21,000 MW constitutes 13% of the total available potential. High head/low power potential adds another 26,000 MW (16% of the total); therefore, low power potential is about 30% of the total available power potential.

Over 90% of available power potential could be realized using conventional turbines, but perhaps in new system configurations. However, nearly two-thirds (66%) of the low head/low power potential (»10% of total available potential) corresponds to technologies (unconventional systems and microhydro) that would require additional turbine and system configuration R&D; although, some units currently exist that could be put into service.

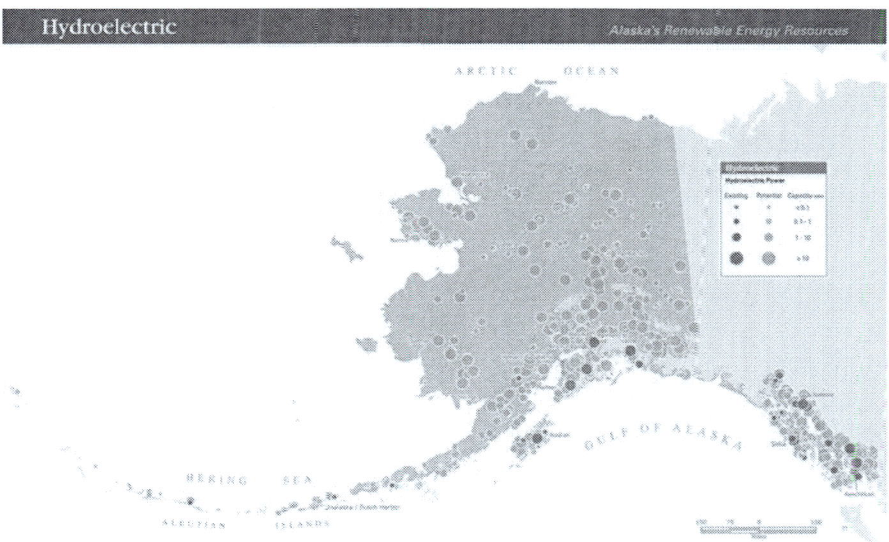


Figure 3. A hydropower-energy resource map for Alaska (source: AEA[4]).

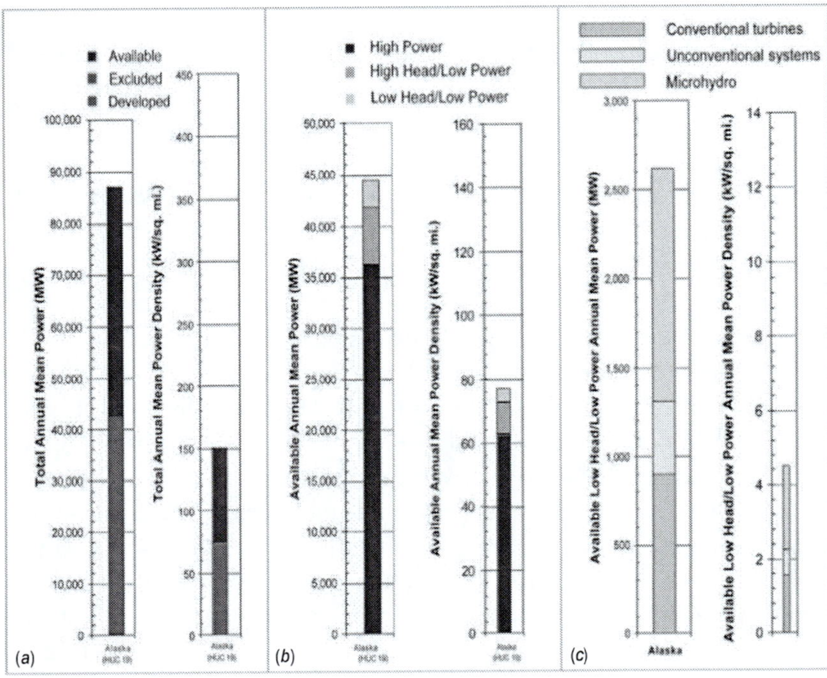


Figure 4. Alaska's hydropower potential. (*a*) Total power potential (*left*) and power potential density (*right*) of water resources in Alaska divided into developed, excluded (by law or treaty), and net constituents. (*b*) The power potential (*left*) and power potential density (*right*) of Alaska's available (from *a*) water energy resources divided into high power, high head/low power, and low head/low power constituents. (*c*) Alaska's available (from *a*) power potential (*left*) and power potential density (*right*) from low head/low power water energy sources divided into conventional turbines, unconventional systems, and microhydro constituents (source: DOE EERE[5]).

## Ocean Energy

With its vast and highly indented coastline, Alaska can also exploit the tidal energy supplied by the Earth-Moon gravitational system. Alaska has many ideal candidate sites where strong tidal currents flow through geographically restricted channels—between islands, between islands and. coastal features, or through undersea canyons or features. While tidal energy is intermittent—it is intermittent in a way that is precisely predictable. It could be regulated in tandem with river hydroelectric power sources to provide continuous reliable power.

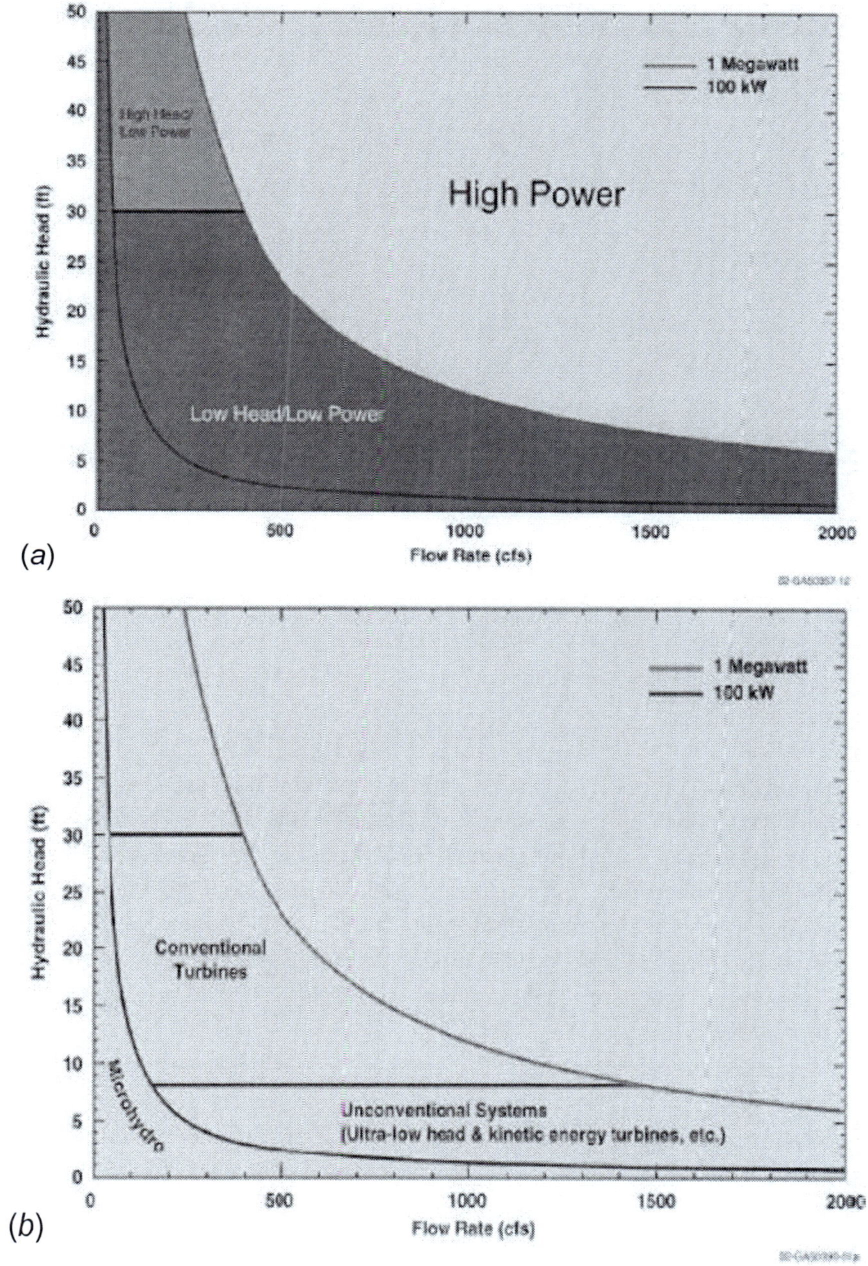


Figure 5. (a) Hydropower energy-collection regimes and (b) hydropower technologies that can capture that energy (source: DOE EERE[5]).

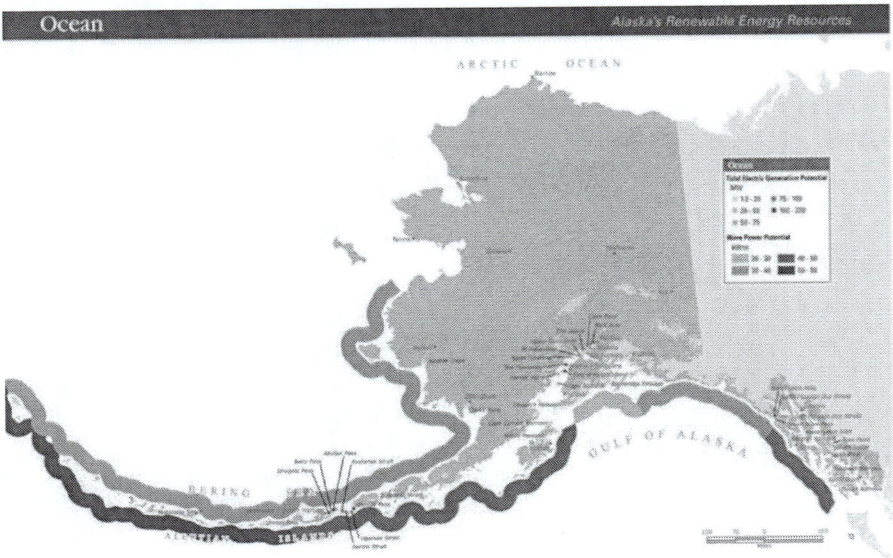


Figure 6. A ocean-energy resource map for Alaska (source: AEA[4]).

## Biomass Energy

According to the AEA, Alaska's most important biomass fuels are wood, sawmill wastes, fish byproducts, and municipal waste. AEA's Biomass Energy Program focuses on developing wood- fired systems that displace fuel oil for heating public facilities, demonstrating fish-oil biodiesel performance, and recovering energy from municipal solid waste. AEA is part of the Pacific Regional Biomass Energy Partnership, a state and federally supported effort that encourages bioenergy development in Alaska, Hawaii, Idaho, Montana, Oregon, and Washington.

Wood remains an important RE source for Alaskans, with over 100,000 cords per year used for residential space heating statewide. Closure of the major pulp mills in Sitka and Ketchikan in the 1990s brought an end to large-scale wood-fired power generation in Alaska; however, recent increases in oil prices have raised interest in using sawdust and wood wastes as fuel for lumber drying, space heating, and small-scale power production. Alaska has also seen renewed interest in converting low-value wood and wood wastes to liquid fuels such as ethanol. Under study in Ketchikan is a wood-waste to ethanol project.

Based on Alaska Wood Energy Development Task Group direction, AEA and the U.S. Forest Service (USFS) Region 10 are providing partial funding for the design of a wood-fired district heating system in Craig. If the project proceeds into construction, the facility will displace up to 36,000 gallons of fuel oil per year with local sawmill waste. AEA and USFS are also working with the Juneau Economic Development Council to conduct reconnaissance level feasibility assessments of wood heated facilities in other communities.

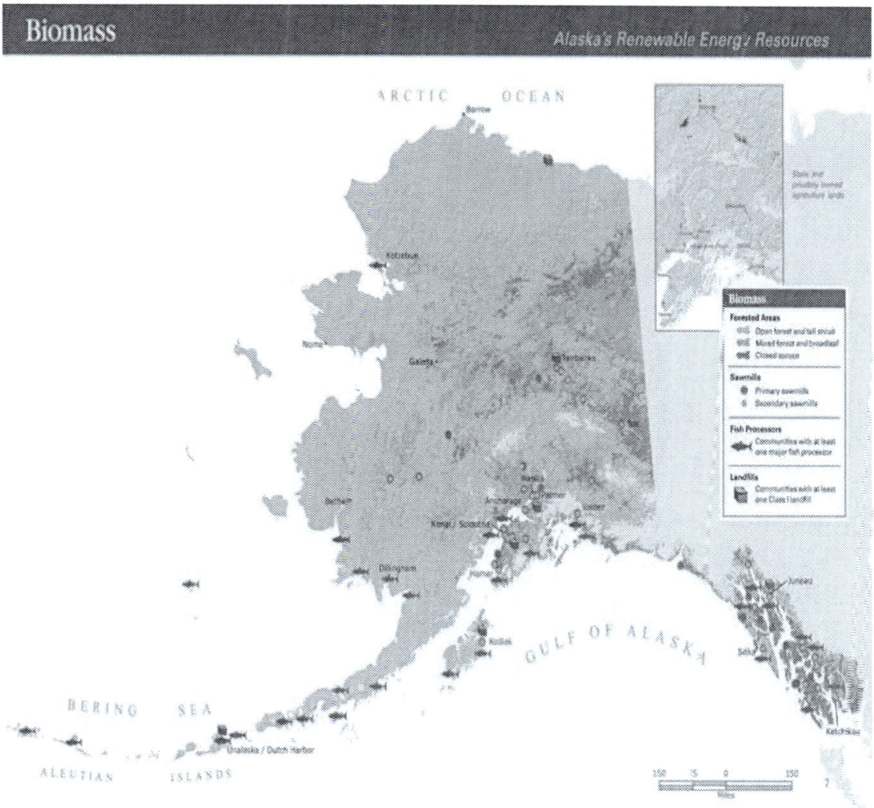


Figure 7. A biomass-energy resource map for Alaska (source: AEA[4]).

### *Fish Oil and Biodiesel*

Shore-based and floating groundfish processors produce approximately 8 million gallons per year of fish oil as a byproduct of fish meal plants. Much of the oil is used in the process as boiler fuel for drying the fish meal or exported to Pacific rim markets for livestock and aquaculture feed supplements and other uses. In 2001, with the assistance of AEA and the Alaska Science and Technology Foundation, processor UniSea Inc. conducted successful tests of raw fish oil/diesel blends in a 2.2 MW 2-cycle Fairbanks Morse engine generator. Since then, the company has expanded the operation and used over two million gallons of 50:50 raw fish oil-diesel blend for power production between July 2002 and June 2004.

Biodiesel is an engine fuel manufactured from renewable sources, such as vegetable oils, recycled cooking greases or oils, or animal fats. Biodiesel is a U.S. EPA-approved substitute manufactured to established industry standards. Currently AEA is working with UAF's Arctic Energy Technology Development Laboratory, Alaska Department of Environmental Conservation, and the National Park Service to test performance of biodiesel in generators at UAF and Denali National Park.

***Municipal Waste***

Alaskans generate approximately 650,000 tons of garbage per year. Currently there is no large- scale recovery of energy from burning unsorted garbage in Alaska. The Sitka Waste-to-Energy facility operated from 1985 to 2000 and provided heat to nearby Sheldon-Jackson College. Fairbanks Memorial Hospital operated a small onsite heat recovery incinerator from 1989 to 2001. Channel Sanitation's Juneau incinerators, under consideration for power production and heat recovery in the 1 990s, halted operation in 2003 after a change in ownership.

Eielson Air Force Base, near Fairbanks, densifies paper separated from the local waste stream and co-fires the 4 $cm^2$ "cubes" at the base's coal-fired power plant. Beginning in 1997 the facility has produced 600–3000 tons per year of "refuse-derived fuel" providing 1% to 1.5% of the base's heat and power load. Co-firing densified paper at local power plants is a component of the AEA- assisted Fairbanks North Star Borough Solid Waste Plan's least-cost alternative. Conventional recycling of paper, around half of the waste stream, is economically marginal at best due to Fairbanks' distance from lower 48 markets.

Energy recovery from Anchorage landfill gas is viable, according to an assessment prepared for the Municipality of Anchorage Solid Waste Services with AEA assistance. The report estimates the landfill will produce an average energy equivalent of ~1.9 million gallons of diesel fuel per year over the next 10 years. The gas could be used to heat nearby military or school facilities, or be converted to 2.5 MW of power, enough to supply 2,500 homes in the rail belt. Over the next 30 years, gas and energy output from the landfill is expected to more than double.

# SCIENCE & TECHNOLOGY RESEARCH EFFORTS

## Geothermal Energy

Worldwide geothermal usage is an estimated 15,000 MW of direct use (heating) and more than 8,000 MW of electricity-generating capacity. This generating capacity is about 0.4% of the world total installed generating capacity. The U.S. is the largest producer of geothermal electricity, followed by the Philippines, Mexico, Indonesia, Italy, Japan, and New Zealand†. Iceland is the largest *per capita* producer and user of geothermal energy.

Sandia is presently supporting DOE EERE Geothermal in working with Iceland. Iceland is moving forward from exploiting the geothermal to heat water, heat most of its buildings, and generate electricity. It has completed one pilot project and begun another to produce hydrogen fuel for fuel-cell vehicles from clean, geothermal energy. They hope to power all land vehicles and their fishing fleet (their main industry) by 2050. They are also beginning to make hydrocarbon fuels from the $CO_2$ that is released from the geothermal wells.

Australia and New Zealand also are focusing considerable efforts on exploiting their geothermal resources. Southern Australia is home to extremely large untapped geothermal resources, but no transmission lines exist to get the electricity to load centers, which are a great distance away.

World-wide geothermal research is vast and highly varied. The governments of the developed and much of the developing world support geothermal research. Most geothermal research in the U.S. is conducted in the national laboratory system and focused on large-scale electrical power generation.[6] The two largest programs are at Sandia (drilling) and Lawrence Berkeley National Laboratory (LBNL) (geoscience). The University of Nevada–Reno hosts the Great Basin Geothermal Energy Center which, in partnership with U.S. industry, conducts research toward establishing geothermal energy as a sustainable, environmentally sound, economically viable energy source in the western U.S. Smaller projects underway under the auspices of the DOE EERE program are

† International Geothermal Association, 2007

**Idaho National Laboratory**

- Mitigation of impact of off-design operation
- Power plant costing methodology
- Enhancement of air-cooled condensers
- Microbiological research
- Continual removal of noncondensable gases for binary power plant condensers
- Pipe coatings
- Geothermal process gas monitors

**National Renewable Energy Laboratory**

- Air-cooled condensers
- Component development for ammonia/water power cycles
- Plant performance enhancement and optimization
- Field-verification of small-scale geothermal power plants
- Direct use field verification
- Heat exchanger field tests

**Lawrence Livermore National Laboratory**

- Co-production of silica and other commodities from geothermal fluids
- Silica scale inhibition

**Brookhaven National Laboratory**

- High-performance coating materials
- Field demonstration and evaluation of lined heat exchanger tubes
- Geothermal silica recovery
- High-temperature polymeric elastomers
- Nondestructive testing of corrosion/erosion damage in piping systems

## Wind Energy

Wind turbines are a mainstay of the RE market. As such, they are a constantly improving technology. Leading research involves transitioning from today's technology of turbine blades made of fiberglass to fiberglass/carbon-fiber construction. In addition, researchers are exploring how to improve the labor-intensive manufacturing process to reduce initial investment costs (and produce competitive U.S.-based jobs through advanced manufacturing techniques such as the DOE/ Industry/State jointly funded project in Iowa by TPI Composites).

- Australia (Australian Wind Energy Assn.),
- Austria,
- Canada (Natural Resources Canada)
- Denmark (Danish Energy Authority),
- The European Wind Energy Assn. (EWEA),
- Finland (The National Technology Agence of Finland),
- Greece (The Ministry of Industry, Energy, & Technology),
- • Germany (Federal Ministry for the Environment, Nature Conservation, & Nuclear Safety),
- Ireland (Irish Energy Center),
- Italy (CESI S.p.A. and ENEA Cassaccia),
- Japan,
- Korea (Ministry of Commerce, Industry, & Energy),
- Mexico (Instituto de Investigaciones Electricas),
- Netherlands (The Netherlands Agency for Energy & the Environment),
- Norway (The Norwegian Water Resources & Energy Directorate and Enova SF),
- Portugal (Departmento de Energìas Renováveis),
- Sweden (Energimyndigheten),
- Spain (Instituto de Energias Renovables of the Centro de Investigación; Energetica Medioambiental y Tecnologica),
- Switzerland (Swiss Federal Office of Energy),
- United Kingdom (Department of Trade & Industry),
- United States (The U.S. Department of Energy)

The effort to allow wind turbines to capture as much of the wind's energy as possible has led to the investigation of innovative concepts such as

- modifying blade cross section design to capture wind energy more efficiently,
- embedding sensors and load-control devices during the manufacturing process to improve blades performance and alleviate fatigue loads, and

- developing computational fluid-dynamics tools to maximize structural and aerodynamic efficiency and fully understand the 3-D flow fields in which the turbines operate.

Many large organizations coordinate wind-energy research. One of the largest is the International Energy Agency (IEA). Signatories of the IEA's Wind Agreement includes:

The *CANMET Energy Technology Centre* (CETC-Ottawa) is studying the effects of cold climate on power generation and operational safety of wind turbines. In the University of Manitoba's wind tunnel, severe icing events are simulated in order to study icing's effects on wind turbines. These experiments will contribute to the development of anti-icing and de-icing technologies to mitigate the effect of Canada's harsh climate on the performance of wind turbines. Also, *Clipper Windpower* has certified its wind turbines for cold climate operation.

Most developed nations and many of the developing nations support wind-energy research. An exhaustive list is not possible in a short paper. Below, we describe the DOE EERE's Wind Energy program which supports a wide variety of research in partnership with major U.S. wind- energy companies. The companies and research institutions identified below could form the core list of researchers invited to a meeting in Alaska to discuss how to encourage exploration of Alaska's wind resources and expansion of its utilization.

By the end of 2006, the U.S. wind industry had become one of the fastest growing utility-scale energy resources in the nation. With a current annual growth rate of 27%, U.S. wind-energy capacity increased from 2,500 MW in 1996 to more than 11,500 MW at the end of 2006. The DOE Wind Program aims to significantly increase wind energy use—increasing and diversifying the domestic energy supply; boosting environmental benefits by avoiding pollutant emissions; and strengthening our infrastructure posture by increasing system reliability while reducing economic effects of fuel price or supply disruptions.

In May 2008, the DOE released a first-of-its kind report that examines the technical feasibility of harnessing wind power to provide up to 20% of the nation's total electricity needs by 2030. The report titled "20 Percent Wind Energy by 2030"[7] identifies requirements to achieve this goal including reducing the cost of wind technologies, siting new transmission infrastructure, and enhancing domestic manufacturing capability. Most notably, the report identifies opportunities for 7.6 cumulative gigatons of $CO_2$ to be avoided by

2030, saving 825 million metric tons in 2030 and every year thereafter if wind energy achieves 20% of the nation's electricity mix. As part of President Bush's Advanced Energy Initiative announced in 2006, clean, secure, and sustainable wind energy has the potential to play an increasingly important role in the nation's long-term energy strategy to make investments today to fundamentally change the way we power U.S. homes and businesses and to help reduce greenhouse gas emissions growth by 2025.

"DOE's wind report is a thorough look at America's wind resource, its industrial capabilities, and future energy prices, and confirms the viability and commercial maturity of wind as a major contributor to America's energy needs, now and in the future," said Assistant Secretary of Energy Efficiency and Renewable Energy for the DOE, Andy Karsner. "To dramatically reduce greenhouse gas emissions and enhance our energy security, clean power generation at the gigawattscale will be necessary, and will require us to take a comprehensive approach to scaling renewable wind power, streamlining siting and permitting processes, and expanding the domestic wind manufacturing base."

Prepared by the DOE and a broad cross section of stakeholders across industry, government, and three of DOE's national laboratories—NREL, LBNL, and Sandia, the report presents an in-depth analysis of the potential for wind in the U.S. and outlines a potential scenario to boost wind electric generation from its current production of 16.8 GW to 304 GW by 2030. For its technical report, DOE also drew on the expertise of the American Wind Energy Association and Black & Veatch engineering consultants and the report reflects input from more than fifty energy organizations and corporations.

The analysis concludes that reaching 20% wind energy will require enhanced transmission infrastructure, streamlined siting and permitting regimes, improved reliability and operability of wind systems, and increased U.S. wind manufacturing capacity.

The U.S. leads the world in new wind installations and has the potential to lead the world in total wind capacity by 2010. Last year, U.S. cumulative wind-energy capacity reached 22,613 MW— with nearly 6,000 MW of wind installed in 2008. Wind contributed to more than 30% of *new* U.S. generation capacity in 2007 (2008 statistics have not yet been released), making it the second largest source of new power generation in the nation—surpassed only by natural gas. The U.S. wind energy industry invested approximately $9B in new generating capacity in 2007, and has experienced a 30% annual growth rate in the last 5 years.

To expand wind energy's contribution to the nation, the DOE Wind Energy Program focuses its research on increasing the technical viability of wind systems and increasing the use of wind power in the marketplace. In the area of technology viability, the program is pursuing large wind technology, distributed wind technology, and supporting research and testing. To increase technology use, the program sponsors research into systems integration and technology acceptance.

### *Large Wind Technology*

The performance goal for the large wind turbine research is to reduce the cost of electricity from large land-based wind systems in Class 4 winds (5.8 m/s at a height of 10 m) to 3.6 ¢/kWh by 2012 and 7 ¢/kWh for offshore systems in Class 6 (6.7 m/s at a height of 10 m) winds by 2014. Wind turbines are currently capable of producing electricity at 5–8 ¢/kWh in the Class 4 wind regimes that are broadly available across the United States, depending on many factors, including project financial structure. The program's strategy is to increase the commercial viability and deployment of wind energy by improving the reliability and performance of existing technology while setting the stage for future wind technologies advanced through applied research and market assessment.

### *Prototype Development*

During the past two decades, the Wind Energy Program has worked with industry to develop a number of prototype technologies, many of which have become commercially viable products. One example is the *GE Wind Energy* 1.5-MW wind turbine. As of June 2008, GE had more than 8,500 of these machines installed worldwide. The design of GE's 1.5-MW machine is based on work conducted with GE and its predecessors (Zond and Enron). Since the early 1990s, the program worked with these companies to test components such as blades, generators, and control systems on the various generations of machines that led to GE's 1.5-MW workhorse. Another project that is demonstrating commercial success is the new 2.5-MW wind turbine manufactured by Clipper Windpower. Clipper produced a prototype of its 2.5-MW Liberty wind turbine in 2005 after only three years of cooperative R&D work with the Wind Energy Program. By August 2008, Clipper had installed 205 of its 2.5-MW Liberty machines. Clipper is targeting another 300-plus wind turbines by year-end 2009 and will expand production further the following year. Clipper Windpower currently has sufficient plant capacity and equipment, a trained workforce, and processes in place to assemble more than 500 Liberty turbines annually, with potential for further capacity extensions.

***Component Development***

The program also works with industry partners to improve the performance and reliability of system components. *Knight & Carver's* Wind Blade Division in National City, California, worked with program researchers at *Sandia National Laboratories* to develop an innovative turbine blade that the company expects to increase energy capture by 5% to 10%. The most distinctive characteristic of the sweep twist adaptive rotor blade is a gently curved tip, which unlike the vast majority of blades in use, is specially designed to take maximum advantage of all wind speeds, including marginal speeds. The blade was tested for endurance at NREL in 2008.

*Distributed Energy Systems* (formerly Northern Power Systems) completed a successful partnership with the program to produce a modular, efficient power electronics package that can be scaled for use in a range of wind turbines, from small to multimegawatt systems. According to the company, the new converter improves wind turbine reliability, energy capture, and grid performance. Tests completed on the converter and the Distributed Energy Systems 1.5-MW direct- drive generator, also developed under the program, demonstrated high-quality power output.

To support the development of more reliable gearboxes, the program has worked with several companies to design and test innovative drivetrain concepts. Clipper Windpower's Liberty wind turbine incorporates a highly innovative multiple-drive-path gearbox that feeds four advanced permanent-magnet generators. *Global Energy Concepts* fabricated a 1.5-MW, single-stage drive-train with a planetary gearbox and a medium-speed (190 rpm), permanent-magnet generator that shows potential for reducing tower-head weight and drivetrain costs. Northern Power Systems constructed a permanent-magnet generator with a novel power converter to allow variable-speed operation. The converter was chosen by the American Wind Energy Association for its 2006 Technical Achievement Award.

*Genesis Corporation* is testing a new tooth form for gearboxes that promises major improvements in power density while reducing the costs of these devices. The company completed the first round of testing with positive results and is now working to refine its design.

***Distributed Wind Energy Technology***

The Wind Energy Program goal for small, distributed wind-energy systems was to reduce the cost of energy to 10–15 ¢/kWh in a Class 3 wind resource from a 2002 baseline of 17–22 ¢/kWh. In 2007, the program met that goal through collaborative efforts with industry partners that included *Southwest Windpower*, *Windward Engineering*, *Abundant Renewable Energy*, and *Wetzel Engineering*, reducing the cost of electricity produced by residential-sized turbines (10 kW or less) to 9.9 ¢/kWh. Working with Distributed Energy Systems, *Composite Engineering*, Global Energy Concepts, *Princeton Power Systems*, and *TIAX*, the program reduced the cost of electricity for business/industrial-sized machines (11–100 kW) to 10.7 ¢/kWh. In 2008, the program refocused its efforts on increasing the market for distributed wind systems to meet a growing demand for community-owned projects and local power generation.

To help industry provide consumers with more small wind turbine systems certified for safety and performance, the program launched an independent small wind test project in 2007. The primary objective of this activity is to test commercially available small wind turbine systems that have a high probability of success in the U.S. market over the next several years. The availability of reliable small wind turbines will support the program goal of increasing the number of small wind turbines installed in the U.S. five-fold by 2015.

The Wind Energy Program has worked with several small wind industry partners to develop and test commercially available award-winning small and mid-sized wind generation systems and components. In 2000, DOE's NREL received and R&D 100 Award for its contribution to the development of the Northern Power Systems (now, Distributed Energy Systems) NorthWind 100 wind turbine. The NorthWind 100 wind turbine is a state-of-the-art wind turbine designed for operation in remote, cold-climate conditions. By the end of 2007, 11 of these turbines had been installed and 10 more were sold and awaiting installation. Since then, they have reconfigured its 100-kW cold weather turbine for agricultural and community applications in temperate climates. The company began testing its new prototype at the NREL in 2007.

Southwest Windpower has been working with the program and researchers at NREL's *National Wind Technology Center* (NWTC) for the past several years to develop a 1.8 kW wind energy generator called the Skystream. In 2006, Southwest Windpower received *Popular Science* magazine's "Best of What's New" award for its new wind generator, and it was recognized by *Time* magazine as one of the "Best Inventions 2006." The new turbine has fully integrated electrical components, costs less, is easier to install, and has quieter operation. Since the company began commercial production of the Skystream in 2007, it has sold more than 1,000 units.

As part of its small wind components research, the Wind Energy Program conducted tests at the NWTC with Windward Engineering on its 4.25 kW Endurance wind turbine developed to demonstrate a furling control system. The company began commercial production of the unit in 2008.

Another component project completed in 2006 was the Princeton Power Systems 50-kW AC-AC converter. This new component has a higher conversion efficiency and produces excellent power quality, uses fewer components, and reduces cost.

***Systems Integration***

The goal of the systems integration activity is to complete program activities addressing electric power market rules, interconnection impacts, operating strategies, and system planning needed for wind energy to compete without disadvantage to serve the nation's energy needs by 2012.

As the need to stabilize national energy resources and electricity demands increases, more utilities are seriously evaluating wind power to provide a portion of their generation mix. At the same time, many utilities are expressing concerns about possible impacts on system operations when greater percentages of wind power are introduced into the electric power system. Their concerns, if not adequately addressed, could limit the development potential of wind power in this country.

***Energy Storage***

A key issue is that large penetrations of wind power will require energy-storage systems such as batteries, flywheels, compressed air, pumped hydro, etc. Appropriately sized energy storage mediates between variable generating sources and variable load demand.

The DOE Office of Science, Basic Energy Sciences (BES) hosted a joint basic-science R&D workshop in energy storage with the DOE's Office of Electricity Delivery and Energy Reliability (OEDER) which has resulted in an Electrical Energy Storage (EES) research program. (OEDER is responsible for the national power grid, technology applications, and applied R&D. BES is responsible for basic-science R&D. Sandia is the lead lab supporting Energy Storage Systems program within OEDER.)

The goal of the OEDER Energy Storage Systems program is to develop advanced energy-storage technologies and systems, in collaboration with industry, to increase the reliability, performance, and competitiveness of electric generation, transmission, and use in utility-tied and off-grid systems. The OEDER Energy Storage Systems program is working with BES to choose the research directions of the EES research program. The BES EES basic-research program sponsored a technology resource document, "Basic Research Needs for Electrical Energy Storage,"[8] which provided the groundwork for and served as a basis to inform the deliberation of basic-research discussions.

The panel on Chemical Energy Storage acknowledged that progressing to the higher energy and power densities required for future batteries will push materials to the edge of stability; yet these devices must be safe and reliable through thousands of rapid charge-discharge cycles. A major challenge for chemical energy storage is developing the ability to store more energy while maintaining stable electrode-electrolyte interfaces. The need to mitigate the volume and structural changes to the active electrode sites accompanying the charge-discharge cycle encourages exploration of nanoscale structures. Recent developments in nanostructured and multifunctional materials were singled out as having the potential to dramatically increase energy capacity and power densities. However, an understanding of nanoscale phenomena is needed to take full advantage of the unique chemistry and physics that can occur at the nanoscale. Further, an urgent need exists to develop a fundamental understanding of the interdependence of the electrolyte and electrode materials, especially with respect to controlling charge transfer from the electrode to the electrolyte. Combining the power of new computational capabilities and *in situ* analytical tools could open up entirely new avenues for designing novel multifunctional nanomaterials with the desired physical and chemical properties, leading to greatly enhanced performance.

The panel on Capacitive Storage recognized that, in general, electrochemical capacitors have higher power densities than batteries, as well as subsecond response times. However, energy storage densities are currently lower than they are for batteries and are insufficient for many applications. As with batteries, the need for higher energy densities requires new materials. Similarly, advances in electrolytes are needed to increase voltage and conductivity while ensuring stability. Understanding how materials store and transport charge at electrode-electrolyte interfaces is critically important and will require a fundamental understanding of charge transfer and transport mechanisms. The capability to synthesize nanostructured electrodes with tailored, high-surfacearea architectures offers the potential for storing multiple charges at a single site, increasing charge density. The addition of surface functionalities could also contribute to high and reproducible charge storage capabilities, as well as rapid charge-discharge functions. The design of new materials with tailored architectures optimized for effective capacitive charge storage will be catalyzed by new computational and analytical tools that can provide the needed foundation for the rational design of these multifunctional materials. These tools will also provide the molecular-level insights required to establish the physical and chemical criteria for attaining higher voltages, higher ionic conductivity, and wide electrochemical and thermal stability in electrolytes.

A third panel identified four cross-cutting research directions that were considered to be critical for meeting future technology needs in EES:

1. advances in characterization;
2. nanostructured materials;
3. innovations in electrolytes; and
4. theory, modeling, and simulation.

Energy from intermittent RE sources can also be stored in mechanical, rather than chemical, form. DOE laboratories are working to improve mechanical energy-storage options with research into flywheel or pumped storage—either compressing air or moving water uphill to reservoirs. In each case, electrical energy produced by the intermittent generator source at a time when it is not needed is converted into mechanical potential energy. When electrical demand rises, the stored potential energy is converted back to electricity and sent to the grid.

### *Grid Integration*

Wind-energy researchers are assisting industry partners with a number of projects aimed at increasing utility understanding of integration issues and confidence in the reliability of new wind- turbine products. The DOE has a wind-turbine reliability program that is working with industry and the utilities to assess and increase the reliability of wind turbines and farms. Information and tools gained from the projects will be distributed through a national outreach effort to investor- owned utilities, electric cooperatives, public power organizations, and energy regulators to encourage the inclusion of wind power in generation portfolios and ensure the continued growth of the wind-energy industry.

The systems integration strategy is to assist regional electric system planning and operations personnel to make informed decisions about the integration of wind energy into their systems. The Wind Energy Program has identified six target research areas.

- Technology characterization and data collection
- Tools and methods development
- Application and implementation
- Integration of wind energy and hydropower technologies
- Wind energy and hydrogen production
- Wind energy and clean water

***Technology Acceptance***

To support the Advanced Energy Initiative objective to expand the use of wind energy, the Wind Energy Program strives to overcome near-term deployment barriers to wind by enhancing public acceptance, promoting supportive public policies, engaging key stakeholders, and addressing siting and environmental issues. In 1999, only 4 states had more than 100 MW of installed wind capacity. By the end of 2008, 21 states had more than 100 MW, with three more approaching that goal. The goal of this project is for 30 states to have 100 MW or more of wind installed by 2010.

The strategy of the Technology Acceptance effort is to build momentum for wind power's use across the U.S. Using a state-focused strategy for its efforts acknowledges the critical role in policymaking and incentive adoption that states have played in wind development to date. The primary program mechanisms for pursuing this activity are the Wind Powering America (WPA) program and the National Wind Coordinating Committee (NWCC).

WPA was established to identify wind-power barriers and options for overcoming them, primarily at the state level. A package of technical assistance and outreach activities is aimed at key user communities—farmers and ranchers, Native Americans, federal facility managers, rural electric cooperatives, and consumer-owned utilities. WPA works with these stakeholders and state and local officials to form state coalitions, or wind working groups, in conjunction with DOE's regional offices, to build the local presence required to accelerate wind power's widespread adoption.

The strategy of the NWCC, a U.S. consensus-based collaborative, is to establish dialogue among key stakeholders, and to support the development of environmentally, economically, and politically sustainable commercial markets for wind power. NWCC members include representatives from electric utilities and support organizations, state legislatures, state utility commissions, consumer advocacy offices, wind equipment suppliers and developers, green-power marketers, environmental organizations, agriculture and economic development organizations, and state and federal agencies.

Many institutional and informational barriers have slowed, and continue to slow, the adoption of wind power. The challenge for technology acceptance is to develop, disseminate, and support an appropriate mix of technical information and general outreach to states where there are strong wind resources yet little public or private momentum on wind energy as an option for development. Another challenge is to bring the wind-energy message to potential users of distributed wind technology. By reaching out to farmers, ranchers, Native Americans, and other state and local stakeholders, WPA can help build a state-level coalition. By building bridges to environmental and regulatory communities, the NWCC helps reduce barriers of interest at the nationally.

The Wind Energy Program's technology acceptance efforts complement the efforts being pursued under other elements of the program. This includes the Systems Integration activity, as both are aimed at reducing undue barriers to wind energy's use. The systems integration work targets some of the more technical barriers, while the technology acceptance efforts tend to address issues associated with state, local, and consumer-owned utility unfamiliarity with the technology and with the need for assistance in overcoming that unfamiliarity. The program is pursuing six specific themes under the technology acceptance activity.

- Outreach to state-based organizations
- Support for rural wind development
- Small-wind-energy outreach
- Institution building through utility partnerships
- Support Native American interest in wind power
- Use of wind power to meet federal loads

***Supporting Research and Testing***

Supporting research and testing provides funding to meet the needs of the large wind speed technology and distributed wind technology research activities by bringing specialized technical expertise, comprehensive design and analysis tools, and unique testing facilities to bear on problems that industry will encounter in bringing new wind energy technology to the marketplace.

The strategy of the supporting research and testing effort is to use the research staffs of the NWTC and Sandia National Laboratories to perform wind-technology-specific research targeted to help industry improve the performance of components and fully integrated turbine systems. To that end, program researchers work closely with DOE and industry to define and prioritize those research activities that address their specific long- and short-term requirements. On occasion, program researchers may also contract with universities and other research organizations for supporting research and testing efforts.

NWTC and Sandia staffs provide extensive design review, analysis, and testing support for industry activities including systems analyses, component blade and drive train tests in NWTC facilities, and validation of turbine prototypes in the field. These activities are closely coupled and directly assist industry in achieving design goals with hardware that will meet international design certification standards. These support activities are performed under the auspices of cooperative research and development agreements (CRADAs) or as part of development subcontracts with industry. The following outlines the supporting research and testing research portfolio.

- Enabling research
- Advanced rotor development
  - Blade development
  - Aerodynamic code development and validation
  - Aeroacoustics research and testing
- Site specific design
  - Inflow characterization
  - Design load specification
- Generator, drive train, and power electronics
- Systems and controls
- Design review and analysis
- Testing support
  - Structural testing
  - Dynamometer testing
  - Field testing

### *A New Federal Agreement*

In June 2008, the DOE signed an agreement with six leading wind turbine manufacturers to find ways to improve turbine design and production methods as the industry attempts to boost its contribution to the nation's electric supply.

The agreement calls for a two-year collaboration to research methods to design and fabricate more reliable turbine components; reduce installation and operating costs; address environmental and technical issues; and to develop turbine certification, workforce and grid connection standards. Signatory companies included

- GE Energy
- Siemens Power Generation
- Vestas Wind Systems
- Clipper Turbine Works
- Suzlon Energy
- Gamesa Corporation

## Hydroelectric Energy

Hydropower is the most established of the RE technologies, providing more than 75% of the electricity generated by RE in the U.S. About 10% of all U.S. electricity is generated by hydropower. Hydropower offers significant advantages over other energy sources: it is a reliable, domestic, renewable resource; it emits no harmful atmospheric emissions or greenhouse gases; and it has intrinsic storage which makes it dispatchable. Hydropower projects also provide other benefits, such as water supply, flood control, irrigation, navigation, and recreation.

The goal for the DOE EERE program's hydropower research is to maintain hydropower as an attractive electricity-producing option for the 21st century and increase electricity generation where available without adverse environmental effects (fish injury/mortality from turbine passage and changes to water quality/levels below dams). These increases in generation will be accomplished through a combination of advanced technology including new hardware (e.g., turbines, hydraulic controls, and facilities) to mitigate for environmental effects and operational improvements at the unit, plant, and reservoir systems levels. To develop advanced technology and operational improvements, the industry relies on aggressive R&D and support programs.

DOE's research promises to increase hydropower generation up to 10% at existing dams, while at the same time improving the environmental quality of U.S. rivers. In order to maintain and expand hydropower's contribution to the nation, the Hydropower Technology Program' research focuses on the two elements of its mission.

- Enhancing the technical viability of hydropower systems through advanced hydropower technology and supporting research and testing.
- Expanding the use of hydropower in the marketplace with systems integration and technology acceptance and systems engineering and analysis.

### *Advanced Hydropower Technology*

The Advanced Hydropower Technology activity supports the development of technologies that will enable existing hydropower projects to generate more electricity with less environmental impact. This will be done by

1. developing new turbine systems that have improved overall performance;
2. developing new methods to optimize hydropower operations at the unit, plant, and reservoir system levels; and
3. conducting research to improve the effectiveness of the environmental mitigation practices required at hydropower projects.

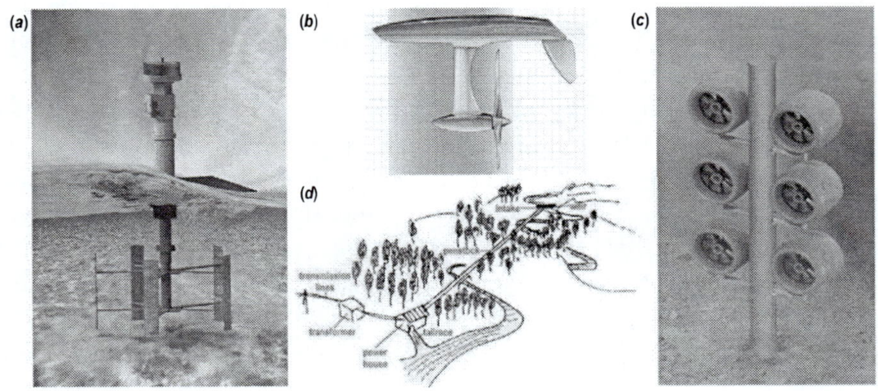


Figure 8. Typical run-of-the-river hydropower technologies are usually either free-flow turbines such as the (a) New Energy Corporation's EnCurrent vertical-axis turbine or (b) Bourne Energy's RiverStarMP 50 which operate in a floating/tethered configuration or (c) Free Flow Power's completely submersible turbine or (d) they require diverting a portion of the river's flow into an engineered system that extracts the energy in a turbine power house and returns the water to the stream.

The main objective of DOE's research into Advanced Hydropower Technology is to develop new system designs and operation modes that will enable both better environmental performance and competitive generation of electricity. While DOE does not own or operate hydropower projects, the products of DOE's research will allow hydropower projects to generate cleaner electricity. DOE-sponsored projects will develop and demonstrate new equipment and operational techniques that will optimize water-use efficiency, increase generation, and improve environmental performance and mitigation practices at existing plants.

When fully adopted, the products of DOE's research will enable up to a 10% increase in the hydropower generation at existing dams, with net benefits to the environmental quality of rivers in the U.S.

Three sets of research projects are being supported under Advanced Hydropower Technology:

- **Large Turbine Testing**. Evaluating a new generation of large turbines in the field to demonstrate that they are commercially viable; compatible with today's environmental standards; and capable of balancing environmental, technical, operational, and cost considerations.
- **Water Use/Operations Optimization**. Developing and demonstrating tools to improve how water is used within hydropower units, plants, and river systems with multiple hydropower facilities, to generate more electricity with less water and greater environmental benefits.
- **Improved Mitigation Practices.** Studying the benefits, costs, and overall effectiveness of environmental mitigation practices and developing guidance that will enable best-available technology to be used in hydro development and operation.

The Hydropower Program is pursuing four specific research activities under supporting research and testing:

- **Biological Design Criteria**. Performing laboratory studies of the biological response of fish to the physical stresses experienced during passage through turbines.
- Computer and Physical Modeling.

- **Instrumentation and Controls**. Developing new instruments and monitoring technologies to measure the physical conditions inside turbines in the field.
- **Environmental Analysis**. Applying advanced computational techniques to describe the full range of hydraulic environments in turbines under different operating conditions.

Two activities are being supported under Systems Integration and Technology Acceptance: hydropower integration with other renewables and outreach activities. Two activities are supported under Supporting Engineering and Analysis: low-head, low-power resource assessment and valuation of hydropower.

***Run-of-the-River Hydropower***

Run-of-the-river hydroelectricity is power generation where the kinetic energy in the natural flow and elevation drop of the water in a river is captured to generate electricity. The best candidates for this type of power generation are rivers with a steady, consistent flow all year round. Run-ofthe-river power projects do not require a large impoundment of water, and so do not disrupt the natural environment as a dam would. The power project can be turbines immersed into the flow of the river's current (see Figure 8a–8c) or installed in a power station on shore (Figure 8d).

In a land-based run-of-the-river power-station, some of the water is diverted from a river (in such a way that aquatic life is kept in the main stream flow) and sent into a pipe called a penstock. The penstock feeds the water downhill to the power station's turbines. Because of the difference in elevation, potential energy from the water up river is transformed into kinetic energy as it flows down the penstock. After going through the generator turbines, the water leaves the power station and is returned to the river (tailrace) without altering the river's total flow or its water levels.

In this system, the turbines are securely mounted in the power house where they are optimized to maximize electricity generation and minimize wear and tear on the machinery. However, the system requires significant infrastructure costs and is difficult to expand without significant additional investment.

In a free-flow run-of-the-river power generating system, an immersed turbine (or an interconnected string of turbines) is either secured to the river bed or suspended in the current from an anchored floating platform. This system requires less infrastructure cost than an engineered system and can be expanded as need demands or financing allows, but the turbines also capture the current's energy less efficiently. Most of these systems and prototypes employ low-RPM turbines that are specially designed to be safe for aquaculture.

Run-of-the-river power stations have little or no capacity for potential-energy storage; they depend on the consistent flow of the river past the power generating turbines. Also, because they cannot store the potential energy of the river, they cannot coordinate electricity generation to match consumer demand or be "timed" to complement other intermittent RE power sources.

## Ocean Energy

### *Ocean Wave Power*

Wave power devices extract energy directly from surface waves or from pressure fluctuations below the surface. RE analysts believe there is enough energy in the ocean waves to provide up to 2 TW of electricity.

Wave power cannot be harnessed everywhere. Wave-power rich areas of the world include the western coasts of Scotland, northern Canada, southern Africa, Australia, and the northeastern and northwestern coasts of the United States. In the Pacific Northwest alone, it is feasible that wave energy could produce 40–70 kW per meter of western coastline. The U.S. west coast is more than 1,600 kilometers long.

Wave energy can be converted into electricity through both offshore and onshore systems. Offshore systems are situated in deep water, typically of more than 40 m. Sophisticated mechanisms use the bobbing motion of the waves to power a pump that creates electricity. Other offshore devices use hoses connected to floats that ride the waves. The rise and fall of the float stretches and relaxes the hose, which pressurizes the water, which, in turn, rotates a turbine.

Specially built seagoing vessels also capture the energy of offshore waves. These floating platforms create electricity by funneling waves through internal turbines and then back into the sea.

Built along shorelines, onshore wave power systems extract the energy in breaking waves. Onshore system technologies include

- **Oscillating Water Column.** The oscillating water column consists of a partially submerged concrete or steel structure that has an opening to the sea below the waterline. It encloses a column of air above a column of water. As waves enter the air column, they cause the water column to rise and fall. This alternately compresses and depressurizes the air column. As the wave retreats, the air is drawn back through the turbine as a result of the reduced air pressure on the ocean side of the turbine.

- **Tapchan.** The tapchan system consists of a tapered channel, which feeds into a reservoir constructed on cliffs above sea level. The narrowing of the channel causes the waves to increase in height as they move toward the cliff face. The waves spill over the walls of the channel into the reservoir and the stored water is then fed through a turbine.
- **Pendulor Device.** The pendulor device consists of a rectangular box, which is open to the sea at one end. A flap is hinged over the opening and the action of the waves causes the flap to swing back and forth. The motion powers a hydraulic pump and a generator.

In general, careful site selection is the key of minimizing the environmental impacts of wave power systems. Wave energy system planners can choose sites that preserve scenic shorefronts. They also can avoid areas where wave energy systems can significantly alter flow patterns of sediment on the ocean floor.

Economically, wave power systems have a hard time competing with traditional power sources. However, the costs to produce wave energy are coming down. Some European experts predict that wave power devices will find lucrative niche markets. Once built, they have low operation and maintenance costs because the energy they capture—motion of the seawater—is free.

### *Ocean Tidal Power*

Typically, ocean or tidal power is harnessed in two ways. The most economical way is with turbines immersed in a fast-flowing tidal stream. Alaska has many ideal candidate sites where strong tidal currents flow through geographically restricted channels—between islands, between islands and coastal features, or through undersea canyons or features. Please see the report "Generation of Tidal Energy" for a detailed description of the current state of tidal-stream turbine technologies and their deployment around the world.

The second way to harness tidal energy is with a constructed barrier that briefly impounds the water and then releases the flow through the part of the structure that contains the generating turbines. Effectively a dam, the traditional structure is called a tidal barrage and is comprised of massive gravity-supported caissons emplaced with great care on a specially prepared site. The tidal barrage site geology must support the extreme weight of caissons that are large enough that their own weight enables them to withstand all of the challenges of the elements.

A newer method borrows construction techniques from the offshore oil & gas industry to build a pile-supported dam. Because the components are supported by piles driven into the underlying geology rather than their own massive weight, they can be much smaller, lighter, and less expensive. Please see the report "Hydroelectric Tidal Wing Dams" for a detailed description of this technology and its potential for application.

While tidal energy is intermittent—it is intermittent in a way that is precisely predictable. It could be regulated in tandem with river hydroelectric power sources to provide continuous reliable power (depending consumer load demands relative to potential energy supply, of course).

## Biomass Energy

The total solar energy absorbed by Earth's atmosphere, oceans, and land masses is ~3,850,000 exajoules (EJ, an exajoule is one quintillion joules) per year.[9] This is more power in one hour than the world used in all of the year 2002.[10] The amount of solar energy reaching the surface of the planet is so vast that in one year it is about twice as much as will ever be obtained from all of the Earth's nonrenewable resources of coal, oil, natural gas, and mined uranium combined.[11] Photosynthesis captures approximately 3,000 EJ per year in biomass.[12] Biomass has been the primary source of energy for most of human history—in addition to being the only source of food.

Biomass is any organic matter, particularly cellulosic or lignocellulosic matter, which is available on a renewable or recurring basis. Plants store solar energy through photosynthesis in cellulose and lig- nin cells. Cellu- lose is a polymer; lignin is the substance, or "glue," that holds the cellulose chain together. When burned, these sugars break down and release energy exo- thermically, giving off heat, steam, and carbon dioxide.

Carbon dioxide is a naturally occurring gas. Plants collect and convert $CO_2$ to aid in photosynthesis. As plants or other matter decompose, or natural fires occur, $CO_2$ is released. Before the discovery and subsequent ever-increasing use of fossil fuels, the $CO_2$ cycle was approximately constant; roughly the same amount that was released through consumption/combustion was sequestered by new plant growth. This approximate $CO_2$ release/sequestration balance has since been lost. In the period since the Industrial Revolution, $CO_2$ levels in the atmosphere have risen from around 150 ppm to 330 ppm, and are expected to double before 2050.

Biomass (wood; crops, e.g., corn, soybeans, or nonfood energy crops) is considered a replenish- able resource—it can be replaced without permanently depleting the Earth's natural resources. Some forms of bioenergy rely on waste from consumer, construction, landfill, industrial, or agricultural sources. These forms are considered renewable because they are produced on a continual basis, and using them is an effective way to put these (already produced) wastes to use.

Biomass can be processed to produce energy in the form of electricity transmitted through the grid, heat for direct use, and transportation fuels. Harnessing this energy through the production of advanced biofuels could meet much of the nation's annual transportation energy needs without producing *net* $CO_2$ emissions that contribute to global climate change.

The biofuels subsidies many Western nations originally supported were founded on the theory that plant-based fuels are carbon neutral, but their equation did not include greenhouse gas (GHG) or other emissions generated from fossil-fuel usage during cultivating, harvesting, or processing the crops into fuel. Nor does it cover the environmental cost of fertilizers and pesticides. The lifecycle energy inputs dictate how much final energy is needed to both satisfy the load and make up for the required energy inputs, while the lifecycle GHG emissions determine the effectiveness of the pathway as a GHG mitigation option. Such factors vary significantly from one biofuel feedstock to another. Increasing evidence shows total emissions and environmental damage from producing many biofuels thought to be "green" often outweigh their lower emissions (when compared with fossil fuels).

Land-use change has transient effects and is more difficult to quantify. Methane emissions from soil bacteria and nitrous oxide emissions from fertilizers change with the crop, but become constant relatively quickly. Changes in soil carbon from cultivating a new crop bring an exchange of carbon with the atmosphere over a period of several years until a new equilibrium is reached.

Once the biomass feedstock is produced, energy inputs and emissions associated with conversion into an energy carrier must also be counted in a consideration of the overall lifecycle from solar energy to energy carrier. These include process heat and any pumping or machinery-assisted mass transfer. The overall efficiency is then calculated by dividing the remaining available energy in the produced energy carrier by the original feedstock's available energy plus the work potential of any energy inputs. Fair comparisons between bioenergy and other conversion pathways can only be performed by considering all lifecycle inputs and emissions.

Governments are responding to these findings. Previously, they sought to encourage biofuel production with blanket incentives and mandates to plant crops and build refineries, as well as tax breaks for the energy (fuel and electricity) producers. Such policies led manufacturers to use the cheapest biomass, rather than biomass that has the fewest harmful effects on the environment. Those policies also produced unintended consequences, like the sale by farmers of food crops for use as biofuel, leading to staple-food price increases.

To address such problems, several countries—including Australia, Britain, France, Germany, The Netherlands, Switzerland, as well as parts of Canada[13]—are revising incentives for farmers, biofuel refiners, and distributors. They are requiring biofuels be certified as "sustainable," on the basis of their emissions reductions and the way the crops are grown, if it is to count toward the annual target of biofuels making up 5.75% of transportation fuel.[13] The manufacturers and sellers will have to *quantify* their fuel's net effect on the environment before being eligible for subsidies or to count toward national biofuel quotas. Very few types of biofuel feedstock sources—like corn stalks—would qualify *automatically* for financial incentives. Food crops—sugar beets, rapeseed, and soy —will fall into a second category, in which producers will have to *prove* that their final biofuel product is environmentally beneficial, on numerous levels.

The next step for many countries is to factor in the broader environmental and social effects of feedstock cultivation, particularly for feedstocks imported from the developing world. The Netherlands announced that it would no longer subsidize the importation of palm oil, its major source of green electricity generation, after investigators showed that the product was grown on Asian plantations created from drained peat land, with disastrous environmental consequences.[13]

Around the globe, bioenergy is most commonly used for heating purposes; applications range in scale from cooking and home heating to industrial process heating. For example, some home-owners burn biomass fuels in wood stoves, while most pulp and paper mills burn wood wastes in large boilers. At these large facilities, bioenergy is often used for electricity generation as well. To generate electricity using bioenergy, the fuel is burned to produce steam that drives a turbine like those at fossil-fueled and nuclear power plants.

Today's bioenergy technology can be used to generate electricity while decreasing harmful emissions. However, individual bioenergy sources can have widely varying environmental effects. Advanced fuels and technologies are being developed that significantly reduce bioenergy emissions, positioning bioenergy as a key contributor to a sustainable energy future worldwide. Much current research focuses on the efficient conversion into fuels of lignocellulosic biomass, the most abundant organic material on the planet. Lignocellulose is a mixture of complex sugars and lignin, a noncarbohydrate polymer that provides strength and structure to plant cell walls. By extracting

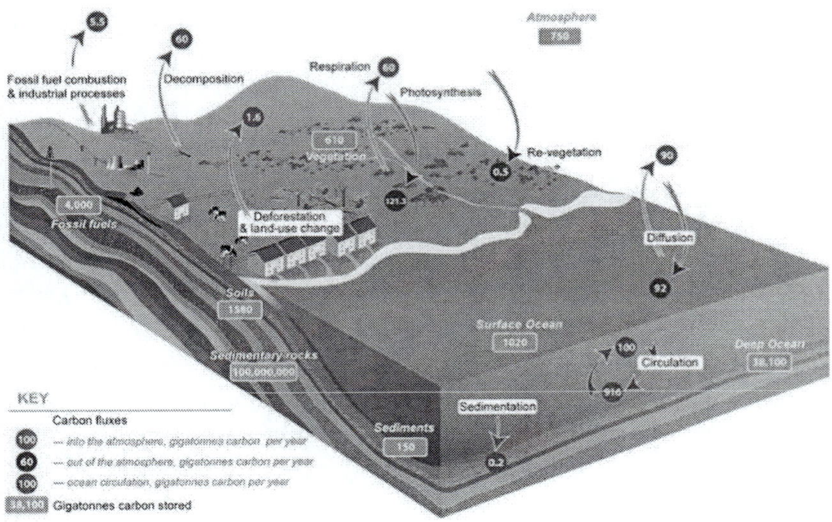


Figure 9. The carbon cycle (source: Cooperative Research Centre for Greenhouse Gas Technologies [Australia/New Zealand]).

simple fermentable sugars from lignocellulose and producing biofuels from them, we can realize the potential of the most energy-efficient and environmentally sustainable fuel crops.[14]

### *Processing Bioenergy Waste Streams*

It must be remembered that biomass energy systems generate waste streams just as conventional fossil energy systems. As with fossil energy systems, the energy to appropriately process the wastes that remain after the energy has been generated from the biomass and that waste's disposal requirements (e.g., landfills) must be taken into account in order to properly measure the success of the RE system with regard to the fossil energy system it replaces.

Japan is a significant example.[15] Its lack of significant domestic energy resources (traditional fossil fuels) in combination with very limited space for landfills has led to their development of an extensive solid-waste-combustion-for-energy industry. Environmental issues related to emissions from the waste-combustion facilities and leaching problems from the generated ash caused the government to investigate and invest in better air-pollution control technologies and methods to stabilize the ash. The Japanese research effort has developed a range of processes including high- temperature gasification (oxygen blown or plasma arc) with ash melting and specific plasma arc systems for melting ash from municipal solid waste combustion facilities.

A Nippon Steel process uses a fixed bed gasifier (not clear if pressurized), with enriched oxygen air injection in the melting section. Coke is added to the municipal solid waste (~50 kg/metric ton municipal solid waste or 5% by weight) input feed which reacts with the oxygen and pyrolytic gases at the air injection and melting region. This is apparently done to help provide energy for full ash melting. Limestone is also added (~5% by weight of input) to provide some pH buffering of the melt. The producer gas is burned in conventional steam boilers from which heat and power are generated. Output materials include granulated slag (90 kg/metric ton input), recyclable iron (10 kg/metric ton input) and fly ash (~30 kg/metric ton input) which is sent to landfill. Mercury and heavy metals present in the waste are found in the fly ash and producer gas—requiring that these streams be managed appropriately before discharge.

As new bioenergy feedstocks and fuels are researched, developed, and demonstrated, system developers must ensure that the total system feedstocks, energy inputs, usable produced energy, and waste streams and waste processing requirements compare favorably with current methods.

### *Bioenergy Research in the U.S.*

In June of 2007, the U.S. DOE chartered and funded ($375M) three bioenergy research centers:

- DOE BioEnergy Science Center led by Oak Ridge National Laboratory,
- Great Lakes BioEnergy Research Center led by the University of Wisconsin–Madison, and
- Joint BioEnergy Institute led by LBNL

to accelerate development of biofuels generated from plants and microbes. The nation must create the right mix of technologies, processes, and expertise to enable environmentally friendly biofuels production—a mission far beyond the scope of any single organization.

Multidisciplinary teams of top scientists from 18 of the nation's leading universities, 7 DOE national laboratories, at least 1 nonprofit organization, and a range of private companies support the three centers. The centers are located in geographically distinct areas and will use different plants both for laboratory research and for improving feedstock crops.

A major focus is on understanding how to re-engineer biological processes to develop new, more efficient methods for converting the cellulose in plant material into ethanol or other biofuels that serve as a substitute for gasoline. This research is critical because future biofuels production will require the use of feedstocks more diverse than corn, including cellulosic material such as agricultural residues, grasses, poplar trees, inedible plants, and nonedible crop portions.

The mission of the centers lies at the frontier between basic and applied science and maintains a focus on bioenergy applications. These centers aim to identify real steps toward practical solutions regarding the challenge of producing renewable, carbon-neutral energy. At the same time, the centers are grounded in basic research, pursuing alternative avenues and a range of high-risk, high-return approaches to finding solutions. To some degree, one key to the centers' success will be their ability to develop the more basic dimensions of their research to a point that can easily transition to applied research.

The centers address inherently interdisciplinary scientific problems requiring scientific expertise and technical capabilities spanning the physical and biological sciences, including genomics, microbial and plant biology, analytical chemistry, computational biology and bioinformatics, and engineering.

The DOE bioenergy research centers are by no means the only avenue for this research in the U.S. Many universities host their own research programs and initiatives. For example, the University of California–Davis has the Bioenergy Research Group, the University of Illinois has the Center for Advanced BioEnergy Research, and the University of Mississippi has the BioEnergy Research Laboratory.

In February 2007, global energy firm BP selected the University of California–Berkeley, in partnership with LBNL and the University of Illinois–Urbana-Champaign, to lead a $500M research effort to develop new sources of energy and reduce the impact of energy consumption on the environment. The funding created the Energy Biosciences Institute, which initially will focus its research on biotechnology to produce biofuels—turning plants and plant materials, including corn, field waste, switchgrass, and algae into transportation fuels.

Below is a short list of other bioenergy initiatives underway in the public and private sectors within the U.S.

- The *American Bioenergy Association* represents ethanol fuel producers, biomass power producers, chemical companies, utilities, farmers, equipment manufacturers, environmental groups, the forest products industry as well as nontraditional allies and partners who support the goals of promoting the economic and environmental benefits of biomass utilization.
- The U.S. Department of Agriculture's *Agricultural Research Service* maintains biofuels research programs for ethanol and biodiesel.
- NREL maintains technical information on biofuels conversion research conducted under DOE supervision from 1980 to the present in its *Biofuels Information Center.*
- The Biomass *Research and Development Initiative* is the multiagency effort to coordinate and accelerate all federal bio-based products and bioenergy R&D.
- The *Northeast Regional Biomass Program* (NRBP) evaluates biomass technologies and fuels and to provide objective, reliable information to consumers and policy leaders. The NRBP carries out its mission through an extensive network of local, state, and national government organizations, and partnerships with private industry.
- The DOE began the *Biomass Power Program* in 1991 to expand the use of RE from biomass.
- The *National Biodiesel Board* (NBB) established itself as the creator of the biodiesel market in the U.S. It is an integral part of the overall exchange of information for the biodiesel commercialization effort, which includes all feedstock providers, government agencies, customers, engine manufacturers, fuel providers, and other interested parties. NBB developed minimum quality standards and a system to register fuel suppliers assuring a high quality fuel supply and consumer confidence.
- The *Renewable Fuels Association* (RFA) is a national trade organization for the U.S. fuel ethanol industry. RFA has been working on behalf of the industry since 1981 to secure a strong

marketplace for ethanol. As the "voice of the ethanol industry," the RFA is dedicated to the continued vitality and growth of ethanol in the fuel marketplace.

Internationally, the IEA (International Energy Agency) hosts *IEA Bioenergy* to improve cooperation and information exchange between countries that have national programs in bioenergy research, development, and deployment.

## Solar Energy

Solar technology research primarily focuses on improving the percent efficiency of the solar collectors. Photovoltaic devices collect energy from roughly 1/3 of the visible-light spectrum. Different substrates collect the energy from different thirds. The most expensive photovoltaics (e.g., the ones on spacecraft and the Mars rovers—called multi-junction photovoltaics) use a combination of these substrates to capture energy from most of the spectrum. This technology is much more efficient, but are far from cost effective for consumer applications. NREL is working to develop substrates that can achieve the effects of multi-junction cells at economical costs.

**Table 2. Solar Energy Research Institutions and Their Areas of Research**

| Name, Location, Type of Institution | Research/Product Type |
|---|---|
| **Florida Solar Energy Center** Cocoa, Florida, USA (trade association, research institution) | Energy-efficient homes and buildings, solar water- heating systems, photovoltaic systems, and solar water- pumping systems |
| **Institute of Energy Conversion** University of Delaware, Newark, Delaware, USA (nonprofit organization, research organization) | Thin-film sola r photovoltaic (PV) cells |
| **National Renewable Energy Laboratory** Golden, Colorado, USA (government research laboratory) | RE research, new technologies, and information |
| Sandia National Laboratories Photovoltaic Program Albuquerque, USA (nonprofit organization, government organization, research organization) | PV cells and PV modules/panels, concentrating solar (PV and thermal), inverters/controllers–balance-of-system components, component testing and evaluation |
| **Renewable and Appropriate Energy Laboratory** University of California, Berkeley, Berkeley, California, USA (research institution) | Solar and RE R&D |
| **Schatz Energy Research Center** Humboldt State University, Arcata, California, USA (nonprofit organiza- tion, research institution) | Fuel-cell test stands, fuel cells, solar-hydrogen power systems, and RE consulting services and feasibility studies |

**Table 2.**

| | |
|---|---|
| **Institut für Solartechnik SPF (Solar Energy Lab)** Rapperswil, Switzerland (research & testing laboratory) | Solar-thermal energy materials, components, collectors, and systems software |
| **Institut für Thermodynamik und Wärmetechnik** Stuttgart, Germany (research institution) | Solar-thermal energy systems |
| **Institut für Solarenergieforschung GmbH** Emmerthal, Germany (research institution) | PV cells and solar-thermal energy systems |
| **Fraunhofer Institute for Solar Energy Systems** Freiburg, Germany (research institution) | PV modules, PV systems, and solar-electric power systems |
| **Ekomation Solar Energy Consultancy** Rotterdam, Netherlands (independent consultancy, research, business intelligence, service) | PV systems, solar water-heating systems, PV modules, solar outdoor-lighting systems, and policy studies |
| **Netherlands Energy Research Foundation** Petten, The Netherlands (nonprofit organization, research institution) | Solar energy and wind energy |
| **Centre for Energy Studies** Dhaka, Bangladesh (research institution) | Research on RE systems and organizing training programs and seminars on energy-related issues |
| **Nimbkar Agricultural Research Institute** Maharashtra, India (nonprofit organization, research institution) | Electric bicycles, wood stoves & furnaces, solar water- heating systems, energy-efficient lighting, biomass energy systems, and alternative fuel vehicles |
| **Botswana Technology Centre** Gaborone, Botswana (manufacturer) | PV fluorescent lights, solar lamps, PV street lighting, PV controllers, compressed soil blocks (using Kgalagadi Sand), PV testing services, solar hearing aid, and passive solar architectural designs |
| **Centro de Investigación en Energía** Morelos, Mexico (research institution, government organization) | PV systems designs for rural applications, solar water heating, low-temperature systems designs for domestic and |
| | industrial applications, solar drying systems for agriculture products, water treatment with high-temp. solar heating systems, thin films, & solar control coatings |
| **New Energy and Industrial Technology Development Organization** Tokyo, Japan (nonprofit organization, research institution, government organization) | PV, solar-thermal, wind, and other new energy technologies |

Also, the BOS (balance of systems) accounts for as much as 50% of the cost of PV systems. Sandia is and has a long standing R&D effort in developing advanced inverters and integrated module designs. A solar-energy information exchange Web site[†] lists most of the leading solar research institutions world wide.

## What are Alaska's Existing Renewable Energy Assets?

### Incentives

Alaska residents and businesses can choose from a wide variety of federal- and state-sponsored incentives to take advantage of RE and energy-conservation technologies.

### Alaska Energy Authority Activities

Currently, AEA's Alternative Energy and Energy Efficiency (AEEE) program manages and funds 33 projects and initiatives totaling $31.5 million in state and federal funding. These projects seek to lower the cost of power and heat to Alaskan communities while maintaining system safety and reliability. AEA's AEEE programs promote

- use of RE resources and local sources of coal and natural gas alternatives to diesel-based power, heat, and fuel production and
- measures to improve efficiency of energy production and end use.

AEA publishes the biennial "Alternative Energy and Energy Efficiency Assistance Plan." This plan describes

- available funding and funding that AEA plans to request for alternative energy and energy efficiency assistance,
- types of assistance that AEA provides and plans to provide, and
- criteria for allocating funds.

[†] http://energy.sourceguides.com/businesses/byP/solar/byB/research/research.shtml

**Table 3. State and Federal RE Investment Incentives for Alaska**

| | State of Alaska | Federal |
|---|---|---|
| *Individual/ Household Incentives* | • Energy Efficiency Interest Rate<br>• Reduction Program<br>• Golden Valley Electric Association<br>• (GVEA)—Builder $ense GVEA—Sustainable Natural Alternative Power (SNAP) Program<br>• Home Energy Rebate Program<br>• Second Mortgage Program for Energy Conservation<br>• Small Building Material Loan | • Residential energy-conservation subsidy<br>• Residential energy-efficiency tax credit<br>• •Residential solar and fuel cell tax credit U.S. DOE alternative fuels data center<br>• Tribal energy program grant<br>• USDA rural energy for America program (grants)<br>• USDA rural energy for America program (loan guarantees)<br>• Federal clean renewable energy bonds<br>Energy-efficient mortgages |
| *Corporate/Municipal Incentives* | • Renewable Energy Grant Recommendation Program<br>• [Homeowners'] Association Loan Program<br>• GVEA—Business $ense<br>• Power Project Loan Fund | • Business energy tax credit<br>• Energy-efficient new homes tax credit for home builders<br>• Renewable electricity production tax credit<br>• Energy-efficient commercial buildings tax deduction<br>• Modified accelerated cost-recovery system + bonus depreciation<br>• Residential energy-conservation subsidy exclusion (corporate exemption) |
| *ProductionIncentives* | • GVEA—SNAP Program | • Renewable energy production incentive |

AEA investments fund projects and initiatives in the following areas: biomass energy, diesel efficiency, end-use efficiency (conservation), geothermal energy, hydroelectric power, ocean and river energy, solar power, transmission and distribution (inter-ties), and wind energy.

## Commercial/Nongovernmental Renewable Energy Activities in Alaska

### *Cold Climate Housing Research Center*

The Cold Climate Housing Research Center (CCHRC)[†] is an industry-based, nonprofit corporation created to facilitate the development, use, and testing of energy-efficient, durable, healthy, and cost-effective building technologies for Alaska and the world's cold-climate regions. Ninety percent of CCHRC's charter members are general contractors from across the state. The Alaska professional building community is highly regarded as a national leader in energy-efficient housing design and construction, boasting the largest per capita builders' association in the nation.

The CCHRC research and testing facility is located on the campus of America's Arctic University at UAF (University of Alaska–Fairbanks). CCHRC and UAF share space and resources to advance the understanding and application of cold-climate housing principles—providing space to work more closely with students, faculty, and researchers at the university.

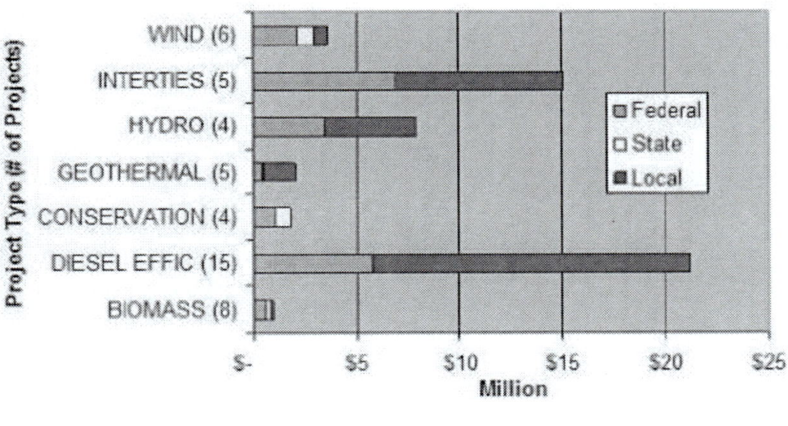


Figure 10. AEA AEEE projects listed by program.

With its new facility, the CCHRC will enter into key research, product testing, and technology- transfer relationships with UAF and building scientists, building industry partners, and home building associations across the state and the nation. The facility provides space for CCHRC and its

† http://www.cchrc.org/

partners to carry out core research and product testing. It also provides classroom, library, meeting, and demonstration space to support technology-transfer agreements. CCHRC's Web site maintains detailed, extensive information on the results of their research into building envelopes, energy systems, foundations and their "green roof."

In October 2007, the Cold Climate Housing Research Center and UAF hosted the first of a series of international conferences on circumpolar housing and community infrastructure issues. This forum presented an opportunity for an international group of builders, architects, building scientists, and planners to discuss common challenges and exchange solutions for producing sustainable, appropriate, durable buildings, and infrastructure in the circumpolar north. The CCHRC plans to continue identifying pressing research needs in the field of northern building and infrastructure design and technology, as well as develop ongoing circumpolar partnerships.

***Alaska Wood Energy Development Task Group***

The Alaska Wood Energy Development Task Group is a coalition that is exploring opportunities to increase the use of wood for energy in Alaska. Since 2005, the task group has solicited statements of interest for thermal wood heat projects in Alaska. From 2005 to present, the task group has received and reviewed 79 statements of interest, selected 42 projects for further study, completed 34 site inspections and field reports, and completed 21 feasibility assessments. Two projects are in the design stage, and 3 projects are installed and operational.

The USDA Forest Service and AEA have been the lead agencies, providing expertise and the bulk of the funding. Juneau Economic Development Council provides the primary "point of contact," resource information, technical assistance, site reconnaissance, and pre-feasibility assessments. The Alaska Wood Energy Development Task Group is made up of the following partners:

- Alaska Energy Authority
- Alaska Village Initiatives
- Bureau of Indian Affairs
- Bureau of Land Management
- Denali Commission
- Division of Forestry, State of Alaska DNR
- Juneau Economic Development Council

- USDA Farm Service Agency
- USDA Rural Development
- National Renewable Energy Laboratory, DOE
- Pacific Northwest Research Station, USDA Forest Service
- State and Private Forestry, USDA Forest Service
- USDA Natural Resources Conservation Service
- University of Alaska–Fairbanks Cooperative Extension Service

***Sustainable Energy Commission of the Alaska Peninsula***

The Sustainable Energy Commission of the Alaska Peninsula (SECAP) is another group that has recently formed. SECAP has installed two grid-tied 10 kW Bergey wind systems at Port Heiden and Pilot Point.

***Pacific Regional Biomass Energy Partnership***

The Pacific Regional Biomass Energy Partnership is one of five regional programs supported by state energy offices or governors associations/regional energy boards/councils and the U.S. DOE. It encourages the development of bioenergy in the states of Alaska, Hawaii, Idaho, Montana, Oregon, and Washington. The DOE Western Regional Office manages the partnership and coordinates with the federal Biomass Program.

The partnership's major purpose is to encourage the deployment of biomass energy technologies. For example, the partnership has worked to develop biodiesel and ethanol production throughout the region. The partnership also has supported biopower development through the testing and demonstration of the anoxic gas flotation process for dairies that use a flush system to handle manure. In addition, the partnership produces and distributes reliable information on potential biomass energy technologies.

The mission of the partnership is to encourage the use and development of biomass energy technologies that are technically feasible and cost effective. They work to provide technology transfer, remove barriers to biomass energy production, and promote its benefits. They provide information and technical assistance to improve the regional environment and economies. The DOE Western Regional Office and each state bring strengths and expertise to the partnership. In addition, the partnership has developed a network of other experts. The range of expertise includes: biodiesel, ethanol, anaerobic digestion and biopower, bioproducts, resource assessments, and policy and siting.

***Renewable Energy Alaska Project***

The Renewable Energy Alaska Project (REAP)[†] is a coalition of small and large electric utilities and utility interests, environmental groups, consumer groups, businesses, and Alaska Native interests with the goal of increasing RE production in Alaska. REAP's mission is "to facilitate the increased development of RE in Alaska through collaboration, education and training, and advocacy." A nonprofit organization, REAP hosted their *4th Annual Renewable Energy Fair* at the Memorial Block at Delaney Park in August 2008. Their Web site promises to offer a complete list of Alaska RE manufacturing, design, and sales firms in the future. Below is a list of commercial firms we were able to assemble during our research on Alaska which exemplifies some of Alaska's RE expertise.

# WHAT CAN SANDIA DO TO HELP?

Because Sandia has ongoing R&D in renewable-, fossil-, and nuclear-energy systems as well as grid integration, we provide a unique capability to perform systems analysis, design, and engineering in the transition of the electric power and transportation infrastructures heavily dependent upon fossil energy sources to a dependence upon more sustainable energy resources. Sandia could contribute to Alaska's developing its RE utilization by applying Sandia's recognized excellence in systems analysis and engineering to evaluating your RE potential. This could be applied on both the domestic and community power-generation scales.

Sandia has an memorandum of understanding (MOU) with AEA. We can support their $100M request for proposals for RE integration through the UAF. Sandia is working on an MOU and CRADA with the UAF Institute of Northern Engineering (INE) to help them develop capabilities to support AEA and the utilities, companies, and others in the energy areas. We are picking a first site, likely to be Kodiak, with INE to begin developing capabilities. Sandia is working with the UAF International Arctic Research Center on climate change modeling and monitoring in conjunction with our atmospheric radiation monitoring site in Barrow.

Sandia can increase its partnership with industry, academic, and government resources within Alaska to help them meet the needs of their

[†] http://www.alaskarenewableenergy.org/

industry/utilities as they change their energy/transportation infrastructures to meet the needs of the state's electricity-consuming customers.

**Table 4. Alaskan Commercial Firms Who Manufacture/Design RE Systems**

| Name, Location, Type of Company | Research/Product Type |
|---|---|
| *Manufacturing* | |
| **Durotest Lighting**, Anchorage Manufacturer, distributor | Energy efficient lighting, light bulbs & tubes |
| **Polar Wire Products, Inc**., Anchorage Manufacturer, retail sales, wholesale supplier | Backup power systems, flooded lead acid batteries, battery cables, DC-to-AC power inverters, hybrid power systems, photovoltaic systems |
| *Design* | |
| **ABS Alaskan, Inc.,** Fairbanks Design, project development services, retail sales, wholesale supplier, installation, maintenance and repair services, recycling services | Wind energy systems (small), DC lighting, RE system batteries, photovoltaic systems, lead-acid batteries, DC-to-AC power invert- ers, remote power systems, AC-to-DC power converters, hybrid energy systems, battery charge controllers |
| **Arctic Technical Services**, Fairbanks System design and retail sales | Photovoltaic systems, air heating systems, air filtering and purification systems, flooded lead-acid batteries, battery connectors, solar tracking systems, efficient oil boilers and heaters, Toyotoves, Quadra-Fire woodstoves, Excell chimney & components, Kyocera solar panels, frames and components, Outback and Trace inverters and controllers, misc. solar components |
| **Remote Power Inc**., Fairbanks System design and service, retail sales, installation, troubleshooting, and repair of off-grid power systems | Outback PS2 power systems; Outback, Trace, Xantrex, and Exelech inverters; photovoltaic modules; wind turbines; propane refrigerators; DC appliances; Staber washing machines; Concorde batteries; Ecofans; meters; and MPPT controllers |
| **Renewable Energy Systems**, Anchorage Consulting, design, engineering, project development services, site survey and assessment services, retail sales, contractor services, construction, installation, maintenance, and repair services | DC-to-AC power inverters, wind turbines (small), deep cycle batteries, wind energy systems (small), and solar energy systems |
| **Solar-Alaska**, Fairbanks System design, wholesale supplier, retail sales, installation, and service | Photovoltaic and hybrid power systems |
| **Susitna Energy Systems**, Anchorage Consulting, design, retail sales, and installation | Photovoltaic modules, wind generators, hydro generators, energy efficiency oil & propane heaters, high-efficiency appliances, TOYO oil heaters, on demand water heaters, satellite telephones, and cell phone boosters |
| **The Mobile Homestead**, Sitka Retail sales | Photovoltaic systems, energy-efficient appliances, battery components, hydro energy system components (small), composting sys- tems, solar water heating systems |

## Geothermal Energy

Sandia's focuses its geothermal systems work on expanding the nation's use of geothermal power. The Geothermal Research Department develops technology aimed at reducing the cost of geothermal energy. The majority of our work focuses on improving key elements of well construction technology: hard-rock drill bits, high-temperature electronics, real-time drilling diagnostics, methods for detecting and controlling lost circulation, and geothermal systems analysis. We work closely with both the geothermal and oil & gas industries and participate in a variety of other drilling-related projects.

### *Drill Bits*

Our most significant contribution has been our research into the development of the polycrystalline diamond compact (PDC) drill bit. Sandia researchers conducted the finite-element modeling analyses, diamond bonding research, cutter testing, bit design/analysis, and laboratory and field testing under CRADAs with industry partners. Sandia catalyzed the development of PDC bits in the drilling industry. They now drill over 60% of borehole footage world wide—a $1.5B industry where the bits also save the industry billions annually because of the extended life of the bits themselves.

### *Telemetry, Electronics, and Diagnostics*

One of the most difficult elements of drilling is knowing what is happening at the point where the bit is cutting the borehole—hundreds or thousands of feet below the surface. To stabilize the drill string (the pipe that drives the bit), mud is pumped into the hole. For some time, the industry has used pulses in that mud to act as a conduit for data—at a rate of 2–5 bits per second. Sandia's geothermal programs dealt with the basic physics issues of wave propagation through the drill pipe and the engineering and applications codes to develop an acoustic telemetry system that has increased the data transmission rate ten-fold (20–50 bits/s).

By definition, geothermal drilling is a high-temperature endeavor—problematic for downhole electronics needed for drilling, logging, and monitoring of geothermal wells. Sandia has become the *de facto* "UL Labs" for testing high-temperature electronic components as well complete downhole tools, seals, batteries, fiber optics, and sensors. Our geothermal scientists work with all major manufacturers to analyze component and equipment failures and provide solutions. We have developed new tools and fabrication methods based on Sandia's capabilities from the weapons programs and supplied prototypes to industry for manufacture. We recently began a joint industry partnership to develop standards and test high-temperature batteries that the oil & gas and geothermal industries need.

Understanding the environment at the bit during the drilling process is important to increasing the rate that holes can be drilled as well increasing the reliability of drilling systems. Sandia developed a diagnostics-while-drilling (DWD) system to understand this drilling process. The DWD system provides the driller and engineers real-time data of the downhole environment and is capable of operating at geothermal temperatures.

### *Drilling Dynamics Systems Modeling*

Dynamic dysfunctions during drilling are among the leading causes of nonproductive time on a drilling site. The drill bit, drill string, and the local geology interact in a complex way that can induce a variety of vibration-related problems—resulting in a low rate of penetration and/or bit and tool failure. If the bit fails, the drill string must be "tripped" to replace it—a process that can exceed a million dollars in some situations.

With a combination of unique software and hardware Sandia has developed a drilling dynamics simulator to evaluate arbitrarily long drill strings in a laboratory environment. Once validated, our drilling dynamics simulator will

- identify deficiencies in drill-bit designs and help manufacturers improve bit and tool performance before a company is committed to expensive field drilling,
- improve the industry's capability for predicting vibrations,
- validate development of hardware and software for downhole tools that combat vibration, and
- develop best practices for mitigating vibration.

### *Controlling Lost Circulation*

During a drilling operation, a drilling fluid is circulated through the drill string and up the borehole to stabilize the borehole, clean cuttings from the hole, and cool and lubricate the bit. As the borehole is extended, it breaches a wide variety of geological formations that exhibit varying degrees of permeability and may contain fractures, faults, or voids. The drilling fluid can be lost from the borehole into the surrounding rock formations. This loss of drilling fluid increases the expense of the drilling operations as the lost fluid volume must be replaced and can cause drilling problems ranging from borehole instabilities to the drill string sticking in the borehole. Sandia has worked on technologies to reduce problems of lost circulation—solutions ranging from the development and testing of high-temperature polymers in borehole grouts to drillable packers (mechanical seals).

### *Water Systems Research*

As water moves through a geothermal system, it has a tendency to coat the inner surfaces of pipes, tanks, and boreholes with a scaly residue. Eventually these system components will become occluded and must be replaced—an expensive process for the above-ground components. The boreholes could only be replaced by drilling new ones. Sandia's water systems researchers are developing a high-capacity, nanometer-scale filtration membrane to reduce the system water's total dissolved solids and remove scale-forming constituents. This new membrane will also facilitate the use of nonpotable, brackish aquifers for the system water for power generation plants—easing the increasing strain on limited potable water sources.

## Wind Energy

### *Materials and Manufacturing*

Wind-turbine blades constitute a significant portion of the cost of a modern, utility-scale, wind turbine. These blades are comprised of relatively low-cost composite materials and current manufacturing processes are very labor-intensive. To facilitate incorporating larger blade designs into new turbines, Sandia studies composite materials and manufacturing processes targeted at developing innovations that will help reduce the nonlinear growth in blade weight. The objective of this effort is to provide innovations in materials, manufacturing processes, and embedded sensor technologies. Sandia scientists are also working with DOE Wind Program, TPI Composites, and the State of Iowa to build a new, highly automated wind-turbine-blade manufacturing facility (GE blades). It may be possible to do something similar in Alaska.

### *Innovative Concepts*

As wind turbines become larger and heavier, blades that incorporate small load-control devices (similar to but smaller than flaps on an airplane wing) and embedded sensors to alleviate fatigue loads offer the potential for significant weight savings. Efforts are focused in three primary areas: (1) analysis of the aerodynamic performance, (2) development of advanced controls, and (3) calculation of the maximum potential cost of energy reductions that can be reasonably achieved through reductions in fatigue loading. These are leading edge sources of R&D for university programs and advanced technology development companies.

### *Aerodynamic Tools and Aeroacoustics*

Sandia continues to develop and utilize computational fluid dynamics codes to improve our understanding of the highly 3-D flow fields under which wind turbine rotors operate. By leveraging the high-performance computing capability of Sandia, these tools provide the necessary information needed to develop the next generation of wind turbine blades that maximize both structural and aerodynamic efficiency. Additionally, Sandia will continue to develop aeroacoustics emission and propagation prediction codes that provide the capability to estimate the noise characteristics of wind turbine rotors. As part of that effort, the aerodynamic performance and acoustic emissions of a rotor with blunt trailing edge airfoils will be compared to those of a similar rotor with conventional airfoils and the effects of various blunt trailing edge treatments on these characteristics will be investigated. This comparison effort is supported by wind tunnel tests to compare the measured noise generation and propagation of a traditional sharp trailing edge airfoil and a structurally efficient flatback airfoil.

### *Design Tools and System Modeling*

Sandia will continue its efforts to develop computational tools to significantly improve the structural and aeroelastic analysis capability available to the wind industry. These analytical capabilities may be used to guide the design of new blades as well as to verify/improve the design of existing blades. The validity of the tools will be demonstrated by continuing a comprehensive design, analysis, build, test, and validation program. A major focus is on better integration of the structural analysis and aeroelastic codes. We hope to reduce design time and produce better and more efficient designs for future wind-turbine hardware.

### *System Performance and Blade Testing*

Full-scale testing of prototype wind turbine blades is vital to asses the structural and aerodynamic performance of advanced concepts. Recently,

Sandia has developed three advanced blade designs which are in the process of being evaluated by a series of structural and aerodynamic tests.

Sandia will continue to conduct both laboratory and field testing of advanced blades in the future, and provide the necessary results to industry to ensure the viability of the unique features of the designs. Additionally, results from the blade testing provide the critical information needed to validate and improve our design codes.

## Biomass/Biofuels

Sandia performs extensive R&D in biofuels. We participate in JBEI and have over a dozen research efforts, funded by LDRD and other sources including significant DARPA funding, to develop fuel-source alternatives to corn and other commercially valuable crops.

In the San Francisco Bay Area, LBNL, Sandia , and Lawrence Livermore National Laboratory; major public and private universities; industry; and federal agencies joined together to create JBEI, the Joint BioEnergy Institute, whose mission is to develop basic science and technology to create environmentally friendly biofuels using plant biomass and microbes. JBEI is centrally situated among JBEI's six institutional partners. It is organized along the lines of a biotech company whose goal is to achieve significant scientific progress within the next five years. JBEI will focus its effort in three key areas: feedstock production, deconstruction, and fuels synthesis. JBEI will apply crosscutting technologies in computational tools, systems and synthetic biology tools, and advanced imaging in a multipronged approach for biomass-to-biofuel solutions in addition to discovery-driven benefits for biohydrogen research, solar-to-fuel initiatives, and broader DOE programs.

Sandia is open to opportunities to form new joint R&D opportunities to expand the research base that is addressing biomass/biofuels technical issues.

## Solar Energy

Sandia has decades of research experience in solar energy which is split between photovoltaics (highly useful in Alaska) and solar concentrators (not commercially viable at Alaska's lattitude). We have long been a leader in photovoltaic research. Our Photovoltaic Laboratories work with the photovoltaic industry and energy users to accelerate the commercial use of photovoltaic energy systems and aid in understanding and improving the performance of those systems.

Our Photovoltaic Systems Evaluation Laboratory (PSEL) provides expertise and test support within several facilities and outdoor sites for evaluating PV and other distributed energy hardware. It includes the following capabilities:

- PV systems evaluations and optimization;
- field testing of arrays and systems in collaboration with system integrators;
- performance testing for modules, arrays, power inverters, charge controllers, and batteries;
- diagnosis of module reliability issues with manufacturing consultation;
- complete performance characterization of PV cells and photo sensors;
- calibration of PV reference cells, reference modules, and solar instruments;
- performance measurements on multiple, interacting distributed energy sources; and
- on-site education and training on the attributes and limitations of photovoltaic and other distributed energy systems.

## REFERENCES

Akpan, I. E., Sasaki, M. & Endoh, N. (2006). "Optimization of Domestic-Size Renewable Energy System Designs Suitable for Cold Climate Regions," *The International Journal for the Japan Society for Mechanical Engineering B: Fluids and Thermal Engineering, 49(4)*, 124 1–52.

Norton, P. & Christensen, C. (2006). "A Cold-Climate Case Study for Affordable Zero Energy Homes," presented at the *Solar 2006 Conference*, Denver Colorado, USA, July 9–13, 2006, conference paper NREL/CP-550-39678, July.

Norton, P., Stafford, B., Carpenter, B., Hancock, C. E., Barker, G., Reeves, P. & Kriescher, P. (2005). "*Affordable High-Performance Homes: The 2002 NREL Denver Habitat for Humanity House, A Cold-Climate Case Study*," NREL technical report NREL/TP-550-3, *1650*, April.

Crimp, P., Devine, M., Lockhard, D., Loewen, R., Rode, C. & Willard, H. (2007). "Renewable Energy Atlas of Alaska: A Guide to Alaska's Clean, Local, and Inexhaustible Energy Resources," *Alaska Energy Authority and Renewable Energy Alaska Project*, July (www.akenergyauthority.org).

Hall, D. G., Cherry, S. J., Reeves, K. S., Lee, R. D., Carroll, G. R., Sommers, G. L. & Verdin, K. L. (2004). "*Water Energy Resources of the United States with Emphasis on Low Head/Low Power Resources*," U.S. Department of Energy report DOE/ID-11111 (April).

Tester, J. W., et al., (2006). "The Future of Geothermal Energy: Impact of Enhanced Geothermal Systems on the United States in the 21st Century," DOE Report INL/EXT0611746, ISBN: 0615134386; 372 pages, (available at: http://geothermal.inel.gov), June.

Lindenberg, S., Smith, B., O'Dell, K., DeMeo, E. & Ram, B. (2008). (Editors), "20% Wind Energy by 2030: Increasing Wind Energy's Contribution to U.S. Electricity Supply," U.S. *Department of Energy report*, DOE/GO-102008-2567, May.

Goodenough, J. B., Abruña, H. D., Buchanan, M. V., Visco, S., Whittingham, M. S., Dunn, B., Gogotsi, Y., Gewirth, A., Nocera, D., Kelley, R. & Vetrano, J. S. (2007). "Basic Research Needs for Electrical Energy Storage," U.S. *Department of Energy report*, July.

Smil, V. (2006). "Energy at the Crossroads," Organization for Economic Co-operation and Development, ISBN 0262194929, [http://www.oecd.org/dataoecd/52/25/36760950.pdf], (17-May), retrieved on 29-Sep-2007, *12*.

"Solar Energy: (2008). A new day dawning?" [http://www.nature.com/nature/journal/v443/n7107/full/ 443019a.html], retrieved 7 August 2008 and Powering the Planet: Chemical challenges in solar energy utilization" [http://web DGN _Powering _Planet.pdf], retrieved 7-Aug.

"Exergy (available energy) Flow Charts" [http://gcep.stanford.edu/research/exergycharts.html].

"Energy conversion by photosynthetic organisms" [http://www.fao.org/docrep/w7241e/ w724 1 e06.htm], Food and Agriculture Organization of the United Nations, retrieved on 25-May-2008.

Elisabeth Rosenthal, (2008). "*Europe, Cutting Biofuel Subsidies, Redirects Aid to Stress Greenest Options*," The New York Times (online, nytimes.com), January, *22*.

Biomass Research and Development Board, "National Biofuels Action Plan," October 2008 (available at http://www1.eere.energy.gov/ biomass

Williams, R. B., Jenkins, B. M. & Nguyen, D. (2003). "Solid Waste Conversion: A Review and Database of Current and Emerging Technologies—Final Report," University of California–Davis, Department of Biological and Agricultural Engineering, California Integrated Waste *Management Board Interagency Agreement,* IWM-C0, *172*, December, 19 and 30.

In: Faces of Alaska
Editors: Mark C. Monteith
ISBN: 978-1-61122-733-8


*Chapter 3*

# THE ALASKA NATURAL GAS PIPELINE: BACKGROUND, STATUS AND ISSUES FOR CONGRESS

***Paul W. Parfomak***

## SUMMARY

Constructing a natural gas pipeline from Alaska's North Slope to the lower-48 states has been a government priority—periodically—for more than four decades. Beginning with the Alaska Natural Gas Transportation Act of 1976, Congress has repeatedly affirmed a national need for an Alaska gas pipeline. In remarks to the press President Obama has likewise described an Alaska gas pipeline as "a project of great potential...as part of a comprehensive energy strategy."

Concerted efforts by Congress, the State of Alaska, and other stakeholders have resulted in new momentum to proceed with an Alaska gas pipeline project. A key step in advancing the pipeline project was the inclusion of $18 billion in federal loan guarantees for such a pipeline in the Alaska Natural Gas Pipeline Act of 2004 (P.L. 108-324). More recent milestones were the State of Alaska's 2008 award to TransCanada Corporation (TransCanada) of a license to build a natural gas pipeline from Prudhoe Bay to the lower-48 states and the concurrent announcement of a competing pipeline proposal, the Denali project, by BP and ConocoPhillips. Both projects plan to conduct open seasons in 2010 soliciting commitments for future gas shipments. If the open seasons yield sufficient interest from gas shippers, either the Denali or TransCanada project could file pipeline siting applications with the Federal Energy Regulatory Commission, Canada's National Energy Board, and other agencies. According to the Denali project's developers, the engineering, scheduling, and cost estimating work for the main pipeline and the associated gas treatment plant are continuing on schedule. TransCanada has been engaged in similar project activities on a comparable timeline.

Notwithstanding recent development progress, many potential obstacles to an Alaska gas pipeline remain at this time, especially the project's economics. A proposal in S. 1462 to increase the pipeline's federal loan guarantee to $30 billion, reported by the Senate Committee on Energy and Natural Resources on July 16, 2009, reflects continuing concerns about the project's economic viability due to low natural gas prices stemming from increased shale gas production. Nonetheless, the TransCanada and Denali developers indicate they are on track to begin construction of one of the proposed projects within the next few years. Proponents maintain that, if an Alaskan gas pipeline begins transporting gas to lower-48 markets by 2020 as anticipated, it would result in reduced U.S. energy prices, increased energy security, and lower U.S. emissions of carbon dioxide. They assert that the project would also create a significant number of jobs and support regional economic development. Like other infrastructure projects in wilderness areas, however, an Alaska gas pipeline would also involve significant environmental costs, many of which have yet to be determined.

Ultimately, the regional effects of any pipeline development on the Alaskan/Canadian environment must be weighed against its economic value, energy security value, and its global benefits in reducing carbon emissions from fossil fuels. To date, the judgment of Congress has favored construction of a pipeline—but ensuring that its public benefits continue to outweigh its costs will likely remain a key oversight challenge for the next decade.

## Introduction

Over forty years ago, large natural gas reserves were discovered at Prudhoe Bay, Alaska. Since that time, Congress has been encouraging the development of those natural gas resources. Principal among its policies has been promoting the construction a natural gas pipeline from the Alaska North

Slope to the lower-48 states. Beginning with the Alaska Natural Gas Transportation Act of 1976,[1] and continuing through the Alaska Natural Gas Pipeline Act of 2004 (P.L. 108-324), Congress has repeatedly affirmed a national need for an Alaska natural gas pipeline.[2] The presidential administrations of Jimmy Carter, Ronald Reagan, and George W. Bush also supported an Alaska gas pipeline project. The Obama administration continues that support. In remarks to the press during his first month in office, President Obama described an Alaska gas pipeline as "a project of great potential...as part of a comprehensive energy strategy."[3]

While it has been on the drawing board for decades, on and off, interest in an Alaska natural gas pipeline has recently revived because of accelerated growth in U.S. natural gas demand, price volatility in the natural gas market, and the increased importation of liquefied natural gas (LNG) from overseas. Moreover, many analysts expect that a national policy of carbon dioxide control could further increase natural gas demand for electric power generation and, possibly, transportation fuel. These factors have led legislators to revisit the potential for an Alaska natural gas pipeline to meet future growth in U.S. demand for natural gas and to restart the process of Alaska gas pipeline development.[4] Important milestones in this activity were Alaska's August 2008 award to TransCanada Corporation (TransCanada) of a license to build a natural gas pipeline from Prudhoe Bay to the lower-48 states and the concurrent announcement of a competing pipeline proposal. To date, however, despite renewed support at both the federal and state levels, construction of an Alaska natural gas pipeline has not begun.

## Issues for Congress

This report provides a brief review of the history of efforts to develop an Alaska natural gas pipeline, including project status, recent developments, and the current project outlook. It also discusses key policy issues related to the construction of the pipeline and its potential role in the context of U.S. energy and climate policy.[5] Specific issues of partricular interest to Congress include the implications for U.S. energy supplies and energy prices of an Alaska gas pipeline, and proposed legislation to raise the federal loan guarantee for the pipeline's construction. Other issues include an Alaska gas pipeline's environmental impacts, its physical security, and its relationship to the proposed Mackenzie Valley pipeline in Canada.

## BACKGROUND

Arctic Alaska has substantial natural gas resources. The U.S. Geological Survey (USGS) estimates that conventional natural gas reserves on Alaska's North Slope potentially exceed 100 trillion cubic feet (Tcf), over four times the total annual gas consumption of the United States.[6] The agency's assessment of undiscovered conventional gas resources across the entire Arctic region concluded that over 1,600 Tcf of additional natural reserves gas remains to be found, with a significant share located under U.S. territory.[7] The USGS also estimates that the North Slope, specifically, may contain up to 158 Tcf of technically recoverable natural gas in the form of methane hydrates.[8] Although methane hydrate resources cannot yet be developed because there are no commercially viable methods to do so, future technologies may make such production economic. Taken together, these vast natural gas resources—both proven and anticipated—in Arctic Alaska have been the motivation for a natural gas pipeline to supply Alaskan natural gas to the lower-48 states.

In 1976, Congress passed the Alaska Natural Gas Transportation Act (ANGTA) to provide for sound decision-making on an Alaska Natural Gas Transportation System (ANGTS). The statute provided for congressional and presidential participation in the pipeline planning process and sought to expedite pipeline construction. The three main policy steps in the process were all completed by the end of 1977. First, the Federal Power Commission recommended (equally) two potential gas transportation options, both overland pipeline proposals along different routes through Alaska and Canada to the lower-48 states.[9] (A third, rejected, option was an overland pipeline across Alaska combined with an LNG export terminal in the port of Valdez.) President Carter then selected the route running along the Alaska Highway in preference to the northern route through the Mackenzie Delta, as shown in Figure 1. Congress approved the president's decision through a joint resolution on November 8, 1977.[10] This may have been the last time the project proceeded according to the original plan. A timeline including passage of ANGTS and other key events in the development of the Alaska gas pipeline project is presented in the Appendix.

## Stalled Development

In the winter of 1977-1978, the United States experienced serious shortages of interstate natural gas supplies due to wellhead gas price controls imposed under a 1954 Supreme Court decision.[11] In response to these delivery problems, Congress passed the Natural Gas Policy Act of 1978 (NGPA), which effectively reversed the court's decision. Congress also passed the Powerplant and Industrial Fuel Use Act of 1978 (PIFUA), which restricted the construction of gas-fired power plants and the use of natural gas in large industrial boilers.[12] In the wake of these statutes and related policies, a natural gas oversupply developed, causing natural gas prices to fall sharply and persistently. At the same time, in an often repeated pattern for major U.S. energy supply projects, cost estimates for the Alaska gas pipeline increased. As a result of these factors, the desirability of the Alaska natural gas transportation system declined. [13] Commercial attention to the Alaska gas pipeline initiative essentially disappeared by the mid-1980s.

Although the full Alaska gas pipeline system made little progress, the southernmost parts of the project did get developed during this period to transport growing natural gas supplies from the Western Canadian sedimentary basin. Producers from Alberta, along with Canadian authorities and U.S. and Canadian pipeline companies, cooperated to pre-build the downstream legs of the ANGTS. The western leg of the ANGTS was the Pacific Gas Transmission pipeline, which began service from Alberta to California in 1 98 1. The eastern leg, the Northern Border Pipeline, went into service from Alberta to the Midwest in 1982 (**Figure 1**).

## Marine Transport Options

In the 1980s, reacting to the lack of progress on the land-based pipeline system, the U.S. Maritime Administration authorized a study of marine options for transporting Alaskan natural gas as LNG aboard specialized tankers to determine whether they might offer U.S. shipbuilding opportunities. Such tankers had been transporting LNG to Japan from natural gas fields in south central Alaska (Cook Inlet) since 1969. The results of the study indicated roughly comparable economics for pipeline and LNG transport to the U.S. west coast. LNG sales to the Pacific Rim generally had greater economic potential, but were viewed as politically untenable because they would increase U.S. energy exports rather than increase domestic supplies.[14]

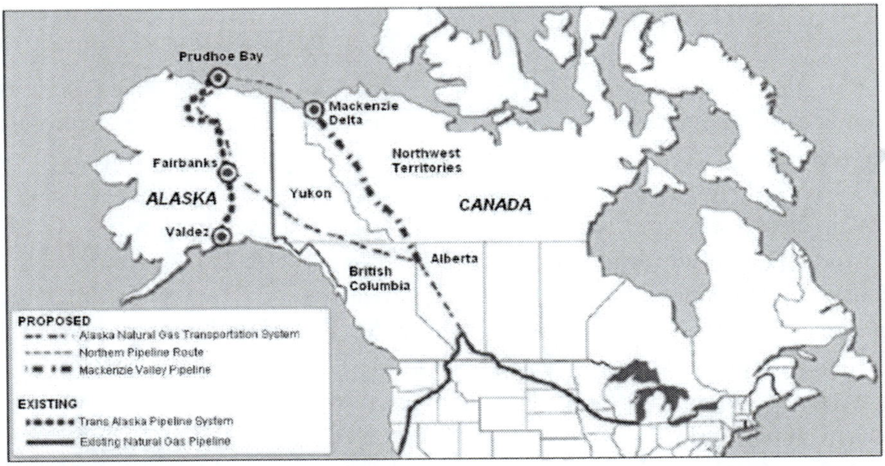


Source: Adapted from CRS Report RL31278, *Arctic National Wildlife Refuge: Background and Issues*, Figure 6, based on Energy Department maps.

Figure 1. Alaska Oil and Natural Gas Pipelines 2009.

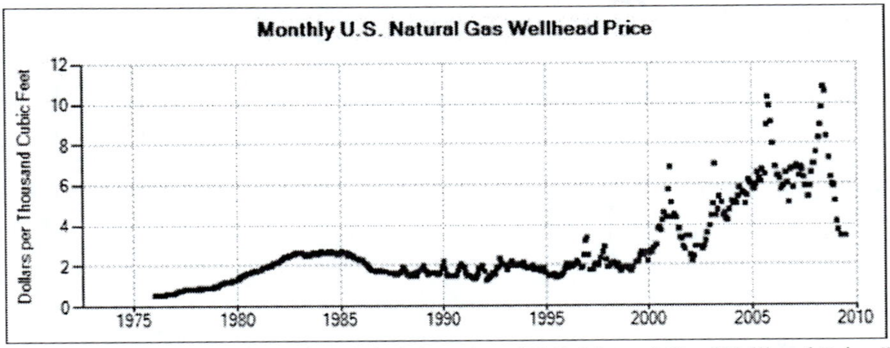


Source: U.S. Energy Information Administration, "U.S. Natural Gas Wellhead Price," Internet database, October 20, 2009, *http://tonto.eia.doe.gov/dnav/ng/hist/n9190us3M.htm*. Prices are in nominal dollars.

Figure 2. Growth and Volatility in U.S. Natural Gas Prices.

## Renewed Interest in the Pipeline

Serious reconsideration of the construction of a natural gas pipeline from the Alaska North Slope began around 2000. This reconsideration was prompted, in large part, by tightening gas supplies to the lower-48 states and corresponding increases in natural gas prices and price volatility as shown in Figure 2.

One important sign of renewed interest in an Alaska pipeline was the recommendation in the 2001 *National Energy Policy* report that the administration "expedite the construction of a pipeline to deliver natural gas to the lower 48 states."[15] Nearly concurrent with the release of this report, President George W. Bush issued Executive Order 13212, which set forth the administration's policy that executive departments and other federal agencies act to expedite projects that would increase the production, transmission, or conservation of energy. [16] To this end a federal interagency task force was established in 2001, co-chaired by the Departments of Energy and State, to identify any impediment to processing an Alaska gas pipeline permit application and to recommend ways to streamline the process.[17]

In 2004, Congress passed the Alaska Natural Gas Pipeline Act (P.L. 108-324, Div. C), a key step in advancing the pipeline project. This act originated in broader national energy legislation that was introduced in 2001, but advanced only slowly through Congress during 2002 and 2003. The Alaska provisions were eventually removed from the broader bill and enacted separately.[18] Among its key provisions, P.L. 108-324:

- clarified that, despite the passage of time since the earlier legislation, the Federal Energy Regulatory Commission (FERC) could still accept, review, and act upon applications for a new pipeline project under the Natural Gas Act or the Alaska Natural Gas Transportation Act;
- created an Office of the Federal Coordinator (OFC) for the issuance by other federal agencies of necessary pipeline permits;
- provided for a pipeline development loan guarantee of as much as $18 billion;
- established guidance for FERC to regulate the pipeline's capacity bidding process so that it would be available to parties beyond the three major North Slope producers— thereby promoting competition in Alaska North Slope natural gas development.[19]

The American Jobs Creation Act of 2004 (P.L. 108-357 § 706) provided a 7-year cost-ofinvestment recovery period for tax purposes (instead of the previously allowed 15-year period) and a designated economic life (class life) of 22 years for the Alaska natural gas pipeline. These provisions were intended by Congress to reduce the cost of capital for the project.[20] The act also provided a tax credit for a North Slope natural gas treatment plant required for the pipeline's operation. Broader, national energy legislation, which passed in 2005 (Energy Policy Act of 2005, P.L. 109-58), also addressed the Alaska natural gas pipeline. P.L. 109-58 required FERC to submit to Congress on a semi-annual basis progress reports about licensing and building the pipeline (§ 1810).[21] In June 2006, consistent with P.L. 108-324 and in accordance with E.O. 3212, fifteen federal agencies with regulatory and other responsibilities related to an Alaska natural gas pipeline signed a memorandum of understanding "to use best efforts to achieve early coordination and compliance with deadlines and procedures established by relevant agencies" in support of the pipeline's development.[22]

### Alaska Gasline Inducement Act (AGIA)

In 2007, Alaska passed the Alaska Gasline Inducement Act (AGIA), seeking to encourage expedited construction of a natural gas pipeline from Alaska's North Slope to gas markets in Alaska and the lower-48 states.[23] AGIA requires a pipeline developer to meet certain requirements advancing the project in exchange for a license providing up to $500 million in matching funds, thereby reducing the developer's financial risks. Among other provisions, the statute requires a developer's commitment to: apply by a specific date for key regulatory approvals, to hold an open season for soliciting bids from North Slope gas producers for future shipments on the pipeline, and to expand the pipeline on rate and tariff terms that would allow gas producers to fully develop Alaska's North Slope natural gas resources.[24] The AGIA also establishes a competitive, public process for the evaluation and selection of pipeline proposals under specific criteria.

## Alaska Gas Pipeline Proposals

Developers have proposed seven major pipeline projects to develop North Slope natural gas reserves since 2007. Five of these proposals were filed under the AGIA process. Two have been proposed outside that process.

## AGIA Proposals

In July 2007, the State of Alaska issued a Request for Applications (RFA) seeking proposals from any interested developer willing to meet the AGIA requirements. The state received competing applications from five developers:

- **TransCanada Pipelines Ltd. (TransCanada)** and Foothills Pipe Lines, Ltd., proposed a pipeline along a route similar to the ANGTS route for the originally approved project. The pipeline would connect with the existing TransCanada Alberta network at the Alberta hub. TransCanada would not construct a North Slope gas treatment plant to process wellhead gas for pipeline transport, although the company stated a willingness to develop such a plant if necessary. The original capacity of this pipeline would be between 4.5 to 5.0 billion cubic feet per day (Bcf/d). According to FERC staff, it would be expandable to 5.9 Bcf/d through the use of greater compression only, and, therefore, at relatively low added cost. [25] ExxonMobil, one of the three major Prudhoe Bay natural gas producers, joined with TransCanada to develop this project in June 2009.[26] The project anticipates an in-service date of 2018.
- **Alaska Gasline Port Authority (AGPA),** a municipal entity (including the City of Valdez, Fairbanks North Star Borough, and North Slope Borough), proposed a pipeline from the North Slope to Valdez, where the gas could be liquefied and shipped as LNG to Asia, the West Coast of the United States, Mexico, or Hawaii. The "All Alaska Gasline/LNG Project" would be constructed and operated by private contractors but publicly financed.
- **Alaska Natural Gasline Development Authority (ANGDA)** proposed a smaller capacity lateral pipeline to link from whatever major North Slope pipeline was selected to South Central Alaska (Anchorage and other locations), making up for declining production there.
- **A China Petroleum and Chemical Corporation (Sinopec)** subsidiary and Little Susitna Construction Company proposed a pipeline from the North Slope to Valdez in the south where the gas would be liquefied and exported to the Pacific Rim as LNG.
- **AEnergia LLC**, a startup company formed by persons with large project engineering experience, proposed a North Slope-to-Alberta pipeline jointly owned by the producers (74%), the State of Alaska (25%), and AEnergia (1%).[27]

Of these applications, former Alaska Governor Sarah Palin determined that only TransCanada's met the AGIA requirements and was complete enough to be evaluated. In May 2008, the governor recommended the TransCanada proposal to the state legislature. The legislature accepted this recommendation and voted to award TransCanada the AGIA license in August 2008.[28] The Sinopec and AEnergia proposals subsequently were not pursued; the other three proposals are still active.

## Other Proposals

In addition to the proposals filed under Alaska's AGIA process, developers have made two alternative proposals to construct North Slope gas pipelines or otherwise bring North Slope natural gas to market.

- **BP and Conoco Philips (Denali),** two of the three major Prudhoe Bay natural gas producers, have jointly proposed a 4.0 Bcf/d pipeline project outside the AGIA process (and, thus, ineligible for the $500 million in state funds). The BP-Conoco Philips "Denali" Pipeline would follow a route similar to that of the TransCanada project to the Alberta hub. The Denali project, however, would include a gas treatment facility and would consider the option of new pipelines through Alberta and to the lower-48 states if deemed necessary. [29] The latest project schedule proposes initial pipeline service in 2018.[30]
- **Enstar Natural Gas Company** has proposed a 690-mile smaller capacity (0.5 Bcf/d) "bullet" pipeline from the North Slope to Southcentral Alaska as an alternative to the ANGDA lateral pipeline proposal. The Enstar pipeline would not depend upon the construction of a larger Alaska natural gas pipeline (either TransCanada's or Denali) to supply natural gas to in-state markets.[31]
- **Alaska Natural Resources To Liquids, LLC** is one of several "gas-to-liquids" (GTL) proposals that would convert North Slope natural gas to liquid hydrocarbon fuels (e.g., diesel and gasoline) and then transport them via a new pipeline or through the existing Trans Alaska Pipeline System (**Figure 1**) to Valdez for marine shipment out of state.[32] While not intended to deliver natural gas to the lower-48 states, such a project would develop Alaska's North Slope gas reserves as an alternative to a natural gas pipeline.

According to the Energy Information Administration, there are no "current" project sponsors actively promoting a GTL project on the North Slope.[33] Of the five Alaska natural gas pipeline projects that are still actively promoted, only the TransCanada, Denali, and AGPA proposals would transport natural gas out of Alaska. The other two proposals, ANGDA and Enstar, involve smaller projects focused exclusively on natural gas supplies to markets within Alaska. Because the latter two proposals are of more state interest than national interest, they are not discussed in the remainder of this report.

## PROJECT ECONOMICS

An Alaska gas pipeline or pipeline/LNG terminal combination transporting natural gas from the North Slope to the lower-48 states would be, by some measures, the largest civilian construction project in the history of North America. When initially announced, the TransCanada and Denali pipeline developers estimated total project costs of $27 billion[34] and $30 billion,[35] respectively. The Energy Information Administration currently assumes total capital costs for a TransCanada or Denali-type pipeline of approximately $28.8 billion and capital costs for a pipeline/LNG project of approximately $35.4 billion.[36] In light of these enormous capital requirements, policy makers and investors alike have paid close attention to key factors affecting the economic viability of the proposed projects, as well as their potential impact on regional and national economic development.

### Natural Gas Price Expectations

Future prices of natural gas will have the greatest influence, by far, on the cost effectiveness of an Alaska natural gas pipeline. According to a 2008 study prepared for the Alaska legislature, the net present value of the TransCanada pipeline would be positive with long-term real natural gas prices above $5.00 per thousand cubic feet (Mcf) and potentially at lower prices.[37] A federal tax credit for the pipeline proposed in 2002 under H.R. 4 implied economic viability of the project at a gas price of $3.25/Mcf ($3.91/Mcf in 2009 dollars).[38] By comparison, the EIA's most recent analysis (which assumes an Alaska gas pipeline is constructed) projects lower-48 wellhead natural gas prices to rise from $4.64/Mcf in 2010 to $8.01/Mcf in 2030.[39] Based on such projections, assuming adequate gas production from North Slope fields, TransCanada expects its pipeline project to be cost-effective, generating $475 billion in total revenue over its operating life for distribution among gas producers and government stakeholders.[40] Assuming similar economics, the Denali project would also likely be cost effective.

Although the EIA's forecast of natural gas prices appears to support the construction of an Alaska gas pipeline, such forecasts are dynamic and can change unexpectedly due to structural changes in natural gas markets, changes in regulation, and general economic conditions. For example, when Congress passed the Alaska Natural Gas Pipeline Act of 2004, the average annual price for U.S. wellhead natural gas was $5.46/Mcf and rising (**Figure 3**). Four years later, in 2008, the average annual gas price increased to $8.07/Mcf. This high price level, however, spurred unexpected growth in the production of "unconventional" natural gas from shale deposits in the lower 48-states.[41] The combination of new shale gas supplies and reduced growth in gas demand due to the recent U.S. economic downturn subsequently caused natural gas prices to plummet. For the first eight months of 2009, wellhead natural gas prices averaged less than $4.00/Mcf. While demand for natural gas will likely rebound as the U.S. economy resumes its growth, independent gas producers assure strong lower-48 domestic natural gas supplies for the foreseeable future—continuing to put downward pressure on gas prices.[42] In such a price environment, there may be ongoing concerns as to the cost-effectiveness of an Alaska gas pipeline; long-term pipeline economics will be a constantly moving target.[43] Nonetheless, Alaska state officials reportedly have stated that, notwithstanding unprecedented shale gas development across North America, the outlook for the pipeline "is extremely healthy from an economic standpoint."[44] The state's Alaska Pipeline Project Report, released in October 2009, likewise states that "the temporary weakness of North American natural gas demand is not a determining factor in establishing the economic viability of an Alaska natural gas pipeline."[45]

## Construction Costs and Cost Uncertainties

TransCanada and Denali have estimated similar costs for their proposed projects—between $28 and $30 billion. Some analysts, however, suggest that these estimates are conservative and that project costs could reach $40 billion.[46] Infrastructure analysts point to numerous examples of cost overruns

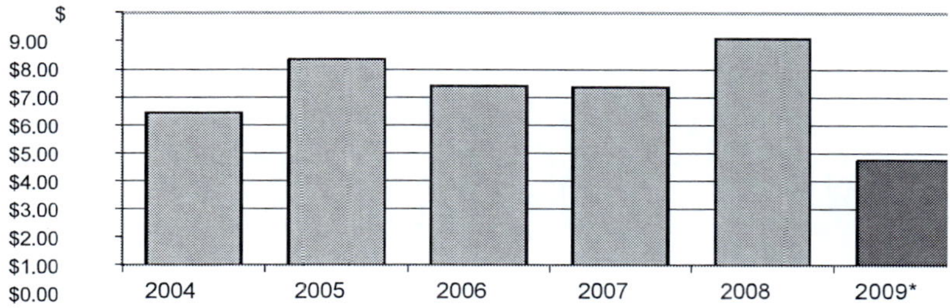


Source: U.S. Energy Information Administration, "U.S. Natural Gas Wellhead Price," Internet database, October 20, 2009, *http://tonto.eia.doe gov/dnav/ng/hist/ n9190us3M.htm.*

*2009 value is year-to-date through August

Figure 3. Annual Average U.S. Wellhead Natural Gas Prices (Nominal dollars per thousand cubic feet).

for other recent multi-billion dollar, decade-long transportation projects, such as Boston's "Big Dig" Central Artery/Tunnel Project, the English Channel Tunnel, and the Oresund Bridge from Denmark to Sweden.[47] They conclude that "large construction cost escalations in transport infrastructure projects are common and exist across different project types, different continents, and different historical periods."[48] While the TransCanada and Denali pipeline developers claim extensive experience managing the construction of major pipeline projects within budget, the scale of an Alaska gas pipeline would be unprecedented, and the risk of cost overruns may be significant. Projected costs for a similar but smaller Arctic natural gas pipeline proposal, the Mackenzie Valley Pipeline (further discussed below), which is also still in development, reportedly doubled between 2005 and 2007.[49]

## Economic Development

The Alaska natural gas pipeline project has important implications for Alaska's economy. By one estimate commissioned by the state in 2006, the net present value of state and local government earnings from the project, including associated North Slope natural gas production, could exceed $29 billion over its lifetime.[50] New natural gas royalty revenue could offset future declines in ongoing oil royalties from mature North Slope oil fields, which have accounted for 85% of the state's unrestricted general fund, on average, since the Trans Alaska Pipeline System began operation.[51] Although state oil revenues set a record in FY2009 due to high global oil prices, oil production in Alaska has actually declined 63% from its peak in 1988.[52]

Development of an Alaska natural gas pipeline to the lower-48 states would also create many new jobs. The Alaska-commissioned analysis concluded in 2006 that the pipeline:

> increases the state's labor force needs by an average of 18,000 direct, indirect and induced workers per year during construction. The project also creates a sustained impact of about 26,000 jobs per year after construction from both pipeline operations and jobs generated by state and local spending of project-related oil and gas revenue.[53]

Throughout the nation, numerous additional jobs would be created in support of, or resulting from, the pipeline's construction. According to one estimate cited in 2004 by Senator Lisa Murkowski, up to 1.1 million jobs nationwide could be created—directly and indirectly—by the pipeline, although details of this analysis are not publicly available.[54] A more recent analysis of public infrastructure spending reported that 21,888 total jobs are created for every $1 billion in new investment in natural gas projects.[55] Based on this ratio, a $28 billion Alaska gas pipeline would create approximately 600,000 new jobs, including jobs in Canada. These jobs include those directly involved in constructing the new project, supplying the project (e.g., line pipe, construction equipment), and induced by the general spending of new pipeline-related employees. Projections of new jobs are highly uncertain, however, particularly for a project as unique as an Alaska gas pipeline, so the actual number of jobs to be created remains somewhat speculative.

In addition to job creation, the presence of a major interstate pipeline from the North Slope to the lower-48 could change the economics of many local energy markets within Alaska. For example, both TransCanada and Denali have agreed to provide up to five natural gas delivery points within Alaska, which could allow local communities near the pipeline right-of-way to provide new natural gas service. Delivery to Anchorage and other Southcentral Alaska markets could also offset declining natural gas supplies from gas fields at Cook Inlet. New or sustained supplies of natural gas to these communities would be an important factor in their economic health.

# Current Project Status

Because the pipeline/LNG proposals for transporting natural gas out of Alaska would involve interstate commerce, they are subject to federal siting approval by FERC under the Natural Gas Act of 1938.[56] Additional state and local approvals for aspects of the project would be required, as well as approvals from Canadian federal, territorial, provincial, and First Nation agencies for the sections of the project running through Canada.

The Denali and TransCanada pipelines, only one of which could be constructed, are proceeding along similar tracks. Both have initiated FERC's pre-filing process, whereby prospective pipeline developers engage with stakeholders—including state, local, and other federal agencies—before filing an application for a Natural Gas Act 7(c) pipeline certificate. Although Denali began the pre-filing process in June 2008, and TransCanada in April 2009, both projects plan to conduct open seasons in 2010 soliciting commitments for future gas shipments. If the open seasons yield sufficient interest from gas shippers, either the Denali or TransCanada project could file pipeline siting applications with FERC, Canada's National Energy Board, and other agencies. According to the Denali's developers, the engineering, scheduling, and cost estimating work for the main pipeline and the associated gas treatment plant are continuing on schedule.[57] TransCanada has been engaged in similar project activities on a comparable timeline.[58] Based on these development schedules, the EIA, Alaska officials, and other stakeholders anticipate that an Alaska natural gas pipeline to the lower-48 states will be in service by approximately 2020.[59]

The TransCanada and Denali pipeline projects are competing and mutually exclusive proposals, each backed by different North Slope producers. To reduce project costs and improve the likelihood of success, some key policy makers have been encouraging the various developers to come together on a single pipeline project. For example, Alaska Governor Sean Parnell has stated, "Four major, international companies all working on an Alaska gas line. To me, that's good news. Now the challenge is to get them working on the same project."[60] To date, the projects are continuing independently, but producer cooperation on a single pipeline is a possibility.

## The LNG Option

The AGPA proposal to build a pipeline from the North Slope to Valdez, and then ship out natural gas on LNG tankers, has not advanced to the detailed planning and project development stage. However, TransCanada has committed to including the option of transporting natural gas for the AGPA's proposed LNG project within its open season for its mainline pipeline from the North Slope to Alberta.[61] According to remarks to Alaska legislators in June 2009, TransCanada has seen "serious interest" among potential shippers for an LNG alternative to its all-land pipeline proposal.[62] Interest in advancing the LNG project has further increased since September 2009, when the AGPA's general counsel, Bill Walker, declared his candidacy for the Alaska governorship with the AGPA gas pipeline/LNG project as the primary issue in his campaign.[63]

# POLICY ISSUES FOR CONGRESS

Congress's last major action to promote an Alaska gas pipeline was passing the Alaska Natural Gas Pipeline Act in 2004. Under that federal statutory framework, as well as provisions in Alaska state law, both the TransCanada and Denali pipeline projects appear to be proceeding on schedule. Since 2004, Congress has been monitoring progress on pipeline development, with only a few active legislative proposals directly addressing these projects (further discussed below). As Congress continues to oversee Alaska gas pipeline activities, several policy issues may attract particular attention. The most important of these is continued uncertainty about whether an Alaska gas pipeline will ultimately be constructed—and, if not, the broader implications for U.S. energy supplies and energy prices. Other important issues are Canada's development of a Mackenzie Valley Pipeline, the federal loan guarantee for the Alaskan pipeline's construction, the environmental impact of the project, and its physical vulnerability to acts of terrorism.

## Alaska Gas Pipeline Development Uncertainty

Many policy makers and developers have high expectations for the contribution of an Alaska natural gas pipeline to U.S. energy supplies. By

2030, the EIA projects an Alaska pipeline carrying 1.99 trillion cubic feet (Tcf) of natural gas annually, which is about 8.5% of total projected U.S. gas supplies.[64] But given the scale, investment requirements, and 10-year time frame required to construct such a pipeline, there continues to be uncertainty regarding its completion. As discussed above, the project faces specific risks deriving from gas prices and pipeline construction costs—but it also faces significant risks involving natural gas production costs, environmental impacts, interest rates, tax policies, and regulatory uncertainty. If, through some combination of these factors, the pipeline's development stalls (again), the anticipated Alaskan natural gas supplies may not become available to lower-48 markets, which may have a ripple effect on the U.S. energy sector.

- **Potentially higher natural gas prices.** Without Alaskan gas supplies, U.S. gas markets would have fewer supply options, and, therefore, might face higher prices. According to the State of Alaska, EIA, and other studies, an Alaska pipeline is expected to be economically viable supplying 1.99 Tcf of natural gas at a market price of $4.00/Mcf to $5.00/Mcf. In the anticipated gas market environment, such a large quantity of relatively low cost gas would help to moderate growth and volatility of natural gas prices in the lower-48 states. This might be especially true if the alternatives for incremental U.S. gas supplies are costlier unconventional gas supplies from North America, or imported LNG—which has a history of price volatility in the international market.
- **Increased LNG imports**. If LNG requirements increase, then U.S. dependence on overseas energy supplies must also increase, counter to a principal energy policy objective of President Obama and many members of Congress. In its current forecast, the EIA projects LNG imports to account for 0.81 Tcf in 2030, or about 3.5% of U.S. natural gas supplies.[65] Absent an Alaska gas pipeline, the United States will have sufficient LNG import capacity in place to double or triple LNG imports if global gas prices warrant it.[66]
- **Carbon control challenges.** Because burning natural gas produces only half as much carbon dioxide as burning coal, and also less carbon dioxide than vehicular fuels like gasoline, many view natural gas as the preferred fuel in the near-term for power plants and motor vehicles under a national policy of carbon control.[67] Not having Alaskan natural gas supplies might, therefore, make it more difficult or costly to reach U.S. carbon reduction targets by substituting natural gas for other fossil fuels in power generation or motor vehicles.[68]

## Canada's Mackenzie Valley Pipeline

Because large sections of a proposed Alaska natural gas pipeline to the lower-48 states would pass through Canada, the involvement of Canadian agencies is essential for the pipeline's success. Canada has cooperated with the United States for decades on a variety of general matters related to the development of an Alaska natural gas pipeline (e.g., the pre-build segments in the early 1980s). However, Canada has been pursuing its own Artic natural gas projects as well. The Canadian project of greatest importance, which could affect an Alaska natural gas pipeline project, is the development of a Mackenzie Valley pipeline.

The Mackenzie River Delta region in the Canadian Arctic contains an estimated 40 Tcf of natural gas.[69] The Canadian government has been interested in developing a pipeline to transport this natural gas to North American markets since it was first discovered in the 1970s in a process that has, in many ways, paralleled the Alaska gas pipeline's development. The main proposal has been a new pipeline from the delta through the Mackenzie River Valley to the existing natural gas pipeline network in Alberta, as illustrated earlier in **Figure 1**. Initially, developers also proposed transporting North Slope gas through a spur pipeline connecting to the proposed Mackenzie pipeline, but the Alaskan connection was rejected by a Canadian government inquiry report in 1974, three years before President Carter also rejected this option.[70] The current configuration of a stand-alone pipeline to Alberta is similar in design to the Alaska gas pipeline proposals, albeit smaller in capacity (1.2 Bcfd) and lower in cost ($15 billion) with a projected in-service date of 2016.[71]

Although there has been significant activity to advance the development of the Mackenzie Valley Pipeline, the project has been controversial due to environmental impacts and escalating costs.[72] According to press reports in October 2009, the Canadian government appeared to be reconsidering the level of financial support it planned to provide the project due to "concerns about the project's price tag."[73] The pipeline's developers maintain that it remains viable, however, and the project continues to receive government support as they proceed with project groundwork and meeting regulatory requirements.[74] Canada's National Energy Board plans to hear final arguments regarding the pipeline's siting application in April 2010, after which it will render a decision as to whether the project will be allowed to proceed.[75]

Because the Mackenzie Valley pipeline would commercialize a major new source of North American natural gas, and would draw on a limited pool of construction resources and materials available for such a project, it has been viewed by some as a direct competitor to an Alaska gas pipeline. [76] For example, the demand for steel and pipe for a Mackenzie Valley project would be significant, and it is not clear that there is adequate, large diameter pipe production capacity in the entire world to supply both Alaska and Mackenzie Valley projects at the same time. Other analysts anticipate that large increases in North American natural gas demand over the next decade will support development of both pipeline projects, and that sufficient construction materials will become available when they are needed. They view the Alaska and Mackenzie Valley pipeline projects as complementary rather than competitive.[77] While the future of natural gas prices and the availability of construction commodities is difficult to predict, because of their physical similarities and links to the same infrastructure, the fates of the Mackenzie and Alaska gas pipelines may remain intertwined.

## Federal Loan Guarantee

As mentioned above, to reduce the financial risks associated with the development of an Alaska natural gas pipeline, the Alaska Natural Gas Pipeline Act of 2004 authorizes the Department of Energy (DOE) to issue up to $18 billion of project loan guarantees for up to 80% of its capital costs and a term of 30 years. The Consolidated Appropriations Act of 2005 (P.L. 108-447) extended the availability of loan guarantees to developers of an LNG project, such as that proposed by AGPA, that would transport natural gas from Southcentral Alaska to West Coast States.[78] DOE solicited comments and information from the public in 2005 in advance of a possible rulemaking concerning the loan guarantee provisions, but has not issued such rules.[79]

The agency reportedly is waiting for the developers to submit more complete project plans before proceeding with specific loan program requirements.[80]

Federal loan guarantees for an Alaska natural gas pipeline have been controversial because they are viewed by some stakeholders, especially supporters of Mackenzie Pipeline development, as an unfair form of subsidy.[81] They also expose the federal government to a significant portion of the financial risk associated with the project. The Congressional Budget Office estimated in 2003 that the loan guarantee under P.L. 108-324 would involve a 10% subsidy and cost $2 billion over the 2010-2013 period, although the agency acknowledged that this estimate was "quite uncertain."[82] Some in

Congress are calling for greater federal loan guarantees for an Alaska gas pipeline. Responding to the recent fall in natural gas prices and tighter credit markets in the wake of the U.S. banking crisis, Senators Bingaman and Murkowski have included provisions in the American Clean Energy Leadership Act of 2009 (S. 1462) that would raise the limit from $18 billion to $30 billion (§ 353), among other provisions. The Senate Committee on Energy and Natural Resources reported S. 1462 on July 16, 2009.

## Environmental Impacts

The addition of significant natural gas resources from Alaska's North Slope to the lower-48 states' fuel supplies is considered by many policy makers to be environmentally beneficial because natural gas produces much lower atmospheric emissions than other fossil fuels (although still far more than renewables, nuclear power, and conservation). Nonetheless the immensity of the roughly 1,750-mile construction project and associated development of North Slope natural gas fields has caused concern about potential environmental effects on land and wildlife. These concerns were heightened by the 2006 shutdown of North Slope oil fields following the discovery of severe corrosion and a small spill from one of BP 's Prudhoe Bay oil pipelines.[83] Broader environmental impacts from existing North Slope oil development and the associated TAPS pipeline have been significant and are well-documented.[84]

Congress has addressed the direct environmental effects of Alaska gas pipeline development by designating FERC as the lead federal agency in preparing an environmental impact statement under the National Environmental Policy Act of 1969 (P.L. 91-190). Under this statute, pipeline permit applicants must prepare certain environmental reports to aid FERC in its preparation of the environmental impact statement (18 C.F.R. § 380.3). The Environmental Protection Agency, as a supporting agency, often assists in the review of the environmental reports and the issuance of the environmental impact statements. In some cases, however, FERC's environmental reviews of major natural gas projects have been challenged in the courts.[85] Given the scale and uniqueness of an Alaska gas pipeline project, future environmental controversies may arise.

One specific environmental issue that faces Alaska gas pipeline development is the potential use of North Slope gas in Alberta's oil (tar) sands industry. North Slope natural gas, if developed, may be used to fuel oil sand operations in northern Alberta. These operations require large quantities of steam to extract crude oil from regional bitumen deposits. Some stakeholders object to the potential use of natural gas for crude oil production on the grounds that it, in their view, consumes a "clean" fuel to produce a "dirty" one, and because oil sands projects can have significant local environmental impacts.[86] The debate over the environmental impacts and potential energy security benefits of Canadian oil sands production is beyond the scope of this report. Nonetheless, given that the primary rationale for federal support of an Alaska natural gas pipeline is to increase lower-48 natural gas supplies, there is the potential for misunderstanding if substantial volumes of North Slope gas are instead diverted to the Canadian oil sands industry as Alaska natural gas begins flowing. This issue may require policy attention.

## Terrorism Risks and Infrastructure Concentration

An Alaska natural gas pipeline would be integral to U.S. energy supply and have vital links to other critical infrastructure, such as power plants, airports, and military bases. While an efficient and fundamentally safe means of transport, such a pipeline would carry volatile material with the potential to cause public injury and environmental damage—either accidentally or intentionally. For these reasons, similar pipelines have been favored targets of terrorists in North America and overseas. Since September 11, 2001, federal warnings about Al Qaeda have mentioned pipelines specifically as potential terror targets in the United States.[87] Congress has responded with substantial initiatives to protect U.S. pipelines from such attacks.[88] Nonetheless, due to its great length and passage through remote areas, an Alaska gas pipeline could still be vulnerable to vandalism and terrorist attack.

Concerns about the security of an Alaska gas pipeline are especially significant because of a recent history of physical attacks on existing pipelines in Alaska and Canada. For example, between October 2008 and July 2009, natural gas pipelines in British Columbia, Canada were bombed six times by unknown perpetrators.[89] In 1999, Vancouver police arrested a man planning to blow up the Trans Alaska Pipeline System (TAPS) for personal profit in oil futures.[90] In 2001, a vandal's attack on TAPS with a high-powered rifle forced a two-day shutdown and caused extensive economic and ecological damage.[91] In January 2006, federal authorities acknowledged the discovery of a detailed posting on a website purportedly linked to Al Qaeda that reportedly encouraged attacks on U.S. pipelines, especially TAPS, using weapons or hidden explosives.[92] In November 2007 a U.S. citizen was convicted of trying to conspire with Al Qaeda to attack TAPS and a major natural gas pipeline in the eastern United States.[93] To date, there have been no known Al Qaeda attacks on TAPS or other U.S. pipelines, but such attacks remain a possibility.

Construction of an Alaska gas pipeline directly alongside TAPS also raises concerns because it increases the concentration of critical U.S. energy infrastructure in the same narrow geographic corridor. When infrastructure is physically concentrated in such a limited geographic area it may be particularly vulnerable to geographic hazards such as natural disasters and certain kinds of terrorist attacks. Whereas a typical geographic disruption is often expected to affect infrastructure in proportion to the size of an affected region, a disruption of concentrated infrastructure could have greatly disproportionate—and national—effects. In 2005, Hurricanes Katrina and Rita demonstrated this kind of geographic impact by disrupting a substantial part of the national U.S. energy and chemical sectors, both heavily concentrated in the Gulf of Mexico. In 2008, Hurricanes Gustav and Ike caused similar disruptions, renewing concerns about geographic vulnerability. Adding 8.5% of U.S. natural gas supplies to a right-of-way that currently delivers nearly 17% of the nation's domestic oil production would be a significant increase in the physical concentration of U.S. energy flows.[94] As development of an Alaska natural gas pipeline continues, physical threats and required security measures may be an important policy consideration.

## CONCLUSION

Constructing an Alaska natural gas pipeline from the North Slope to the lower-48 states has been a government priority—on and off and on again—for more than four decades. Concerted efforts by Congress, the State of Alaska, and other stakeholders have resulted in new momentum to proceed with the project. Many potential obstacles remain at this time, especially the project's economics. The proposal in S. 1462 to increase the pipeline's federal loan guarantee reflects continuing concerns about the project's economic viability. Nonetheless, the developers project the start of construction within the next few years. If the pipeline begins transporting gas to lower-48 markets as anticipated by 2020, it could have an impact on U.S energy prices, energy security, and U.S. emissions of carbon dioxide. The project would also create a significant number of jobs and support regional economic development. Like other infrastructure projects in wilderness areas, however, an Alaska gas pipeline would also involve significant environmental costs, many of which have yet to be determined.

Ultimately, the regional effects of the pipeline's development on the Alaskan/Canadian environment must be weighed against its economic value, energy security value, and its global benefits in reducing carbon emissions from fossil fuels. To date, the judgment of Congress has favored the pipeline—but ensuring that its public benefits continue to outweigh its costs will likely remain a key oversight challenge for the next decade. If the balance tips the other way— either in the eyes of developers or the federal government—and the Alaska gas pipeline is not constructed, Congress may need to redouble its support of other energy initiatives that could fill the substantial expectations unmet by this project.

## APPENDIX

**Table 1. Key Events in Alaska Natural Gas Pipeline Development**

| | |
|---|---|
| 1968 | Prudhoe Bay oil and gas discovered |
| 1969 | United States begins export of LNG to Japan from south central Alaska (Cook Inlet) |
| 1976 | Alaska Natural Gas Transportation Act (ANGTA) passed, P.L. 94-586 |
| 1977 | Presidential Decision and FPC Report to Congress on ANGTS |
| 1977 | FERC (successor to FPC) issues conditional certificate for pipeline |
| 1978 | Trans-Alaskan Pipeline System (TAPS) oil pipeline into service |
| 1981 | "Western leg" of Alaska gas pipeline (Pacific Gas Transmission) into service |
| 1982 | "Eastern leg" (Northern Border Pipeline) into service |
| 1983 | Maritime Administration study of marine alternatives to ANGTS pipeline released |
| 2001 | Alaska Natural Gas Interagency Task Force established |
| 2004 | Alaska Natural Gas Pipeline Act passed, P.L. 108-324, Division C |
| 2006 | New governor announces Alaska Gasline Inducement Act (AGIA) initiative |
| 2007 | Five proposals submitted for AGIA consideration |
| 2008 | Governor determines one AGIA proposal meets AGIA criteria |
| 2008 | Conoco Phillips and BP announce the Denali Project as an alternative to an AGIA project |
| 2008 | Alaska legislature approves governor's AGIA recommendation and it becomes law |
| 2009 | American Clean Energy Leadership Act of American Clean Energy 2009 (S. 1462) proposed to raise loan guarantee |

## End Notes

[1] 15 U.S.C. § 719 et seq.

[2] For example, see P.L. 108-324 § 103(b)(2)(A) "a public need exists to construct and operate the proposed Alaska natural gas transportation project."

[3] Erika Bolstad, "Obama Calls Alaska Gas Pipeline Promising," *Anchorage Daily News*, February 11, 2009.

[4] U.S. Energy Information Administration, *Natural Gas Monthly*, August 2009, Table 2. Full initial pipeline capacity is 4.5 billion cubic feet per day.

[5] Portions of this report previously appeared in CRS Report RL34671, *The Alaska Natural Gas Pipeline: Status and Current Policy Issues*, by William F. Hederman.

[6] David W. Houseknecht, U.S. Geological Survey, *Conventional Natural Gas Resource Potential, Alaska North Slope*, Open File Report 2004-1440, December 13, 2004; U.S. Energy Information Administration, *Annual Energy Outlook 2009*, DOE/EIA-0383(2009), March 2009, p. 109.

[7] U.S. Geological Survey, *USGS Arctic Oil and Gas Report*, Fact Sheet, July 2008.

[8] U.S. Geological Survey, *Gas Hydrate Resource Assessment: North Slope, Alaska*, Fact Sheet, October 2008

[9] The Federal Power Commission (FPC) was the predecessor of the Federal Energy Regulatory Commission (FERC).

[10] H.R. J. Res. 621, P.L. 95-158, 91 Stat. 1268, 95th Cong., 1st Sess., 1977.

[11] Phillips Petroleum Co. v. Wisconsin, 347 U.S. 672

[12] P.L. 95-62 and P.L. 95-620.

[13] See discussion in D. Fried and W. Hederman, "The Benefits of an Alaska Natural Gas Pipeline," *The Energy Journal*, Vol. 2, No. 1, January 1981, p. 22.

[14] W. F. Hederman, "A Review of Marine Systems Use in Developing Alaska Natural Gas," SPE 11294, *SPE Hydrocarbon Economics and Evaluation Symposium*, March 2, 1983, Dallas, TX.

[15] National Energy Policy Development Group, *National Energy Policy*, Office of the Vice President, May 16, 2001, p. 7-18.

[16] E.O. 13212, "Actions to Expedite Energy-Related Projects," 66 F.R. 28357, signed by President Bush on May 18, 2001.

[17] U.S. Department of Energy, National Energy Technology Laboratory, *Delivering Alaskan North Slop Gas To Market*, June 2004, p. 8, *http://www.netl.doe.gov/publications/factsheets/policy/Policy003.pdf*. Other task force members included the Department of the Interior (Bureau of Land Management and Minerals Management Service), the Department of Transportation (Office of Pipeline Safety), and the Federal Energy Regulatory Commission.

[18] State of Alaska, Department of Revenue, *The Alaska Natural Gas Pipeline Act of 2004*, undated, p. 1, http://www.revenue

[19] The FERC issued a final rule on the open season matter on February 9, 2005 (FERC Order No. 2005).

[20] U.S. Congress, Joint Committee on Taxation, *General Explanation Of Tax Legislation Enacted In The 108th Congress*, Joint committee print, JCS-5-05, May 2005, p. 330.

[21] P.L. 109-58, 119 Stat. 594(2005), 42 U.S.C. Section 15801 *et seq.*

[22] U.S. Dept. of Agriculture, et al. ,*Memorandum of Understanding Related to an Alaska Natural Gas Transportation Project*, June 2006, p. 3, *http://www.usace.army.mil/CECW/Documents/cecwo/reg/mou/Alaska_MOU.PDF.*

[23] AS 43.90 et seq.

[24]State of Alaska, "Frequently Asked Questions Regarding the Alaska Gasline Inducement Act ("AGIA") and the AGIA Process," November 8, 2008, p. 1, *http://www.gov.state.ak.us/agia/agia_faqs_11808.pdf.*

[25] Federal Energy Regulatory Commission, *Sixth Report to Congress on Progress Made in Licensing and Constructing the Alaska Gas Pipeline*, August 29, 2008, p. 8.

[26]TransCanada Corp., "TransCanada and ExxonMobil to Work Together on Alaska Pipeline Project," Press release, June 11, 2009.

[27] Federal Energy Regulatory Commission, *Fifth Report to Congress on Progress Made in Licensing and Constructing the Alaska Gas Pipeline*, February 19, 2008, pp. 2-3.

[28] Alaska HB 3001, August 1, 2008.

[29] BP and ConocoPhillips, "BP and ConocoPhillips Join on the Alaska Gas Pipeline," Joint press release, April 8, 2008.

[30] BP and ConocoPhillips, "Denali - The Alaska Gas Pipeline Project," Web page, October 15, 2009, http://www.denalipipeline.com/overview.php.

[31] Fairbanks Economic Development Corp., *In-State Gas Pipeline Supply Options Study*, February 5, 2009, p. 39.

[32] Alaska Natural Resources To Liquids, LLC, "A Legacy Decision for Alaska," Presentation to the Alaska State Legislature, Joint Legislative Budget and Audit Committee, June 20, 2008; Tim Bradner, "Natural Gas: Exxon, BP, Independent All Planning Gas-To-Liquids Projects," *Alaska Journal of Commerce*, April 16, 2000.

[33] U.S. Energy Information Administration, *Annual Energy Outlook 2009*, DOE/EIA-0383(2009), March 2009, p. 41.

[34] TransCanada Corp., *Application for License: Alaska Gasline Inducement Act*, November 30, 2007, p. 2.5-2. 2007 dollars.

[35] Wesley Loy, "BP, Conoco Join Forces to Pursue Gas Pipeline," *Anchorage Daily News*, April 9, 2008. 2008 dollars.

[36] U.S. Energy Information Administration, March 2009, p. 39. Costs are adjusted to 2009 dollars. A gas-to-liquids option is assumed to have capital costs of approximately $60 billion.

[37] Black & Veatch, *Net Present Value (NPV) Analysis*, Presentation to the Alaska State Legislature, Joint Legislative Budget and Audit Committee, June 18, 2008, p. 18, http://lba.legis.state.ak.us/proposals/doclog/2008-06-18black_veatch_npv_spec_sessjnt_comm.pdf.

[38] H.R. 4, Energy Policy Act of 2002, Engrossed Amendment as Agreed to by Senate, Section 2503. The provision was not enacted.

[39] U.S. Energy Information Administration, *An Updated Annual Energy Outlook 2009 Reference Case*, SR/OIAF/2009- 03, April 2009, p. 44. Prices are in 2007 dollars.

[40] Leslie Ferron-Jones, TransCanada Corp., *Projects Impacting Western Markets*, Presentation to the California Energy Commission, 2009 Integrated Energy Policy Report Workshop Sacramento, CA, May 14, 2009, p. 8, *http://www.energy.ca.gov/2009_energypolicy/documents/2009-05-14_workshop/presentations/*13_TransCanada_Ferron_Jones.pdf.

[41] U.S. Energy Information Administration, *Is U.S. Natural Gas Production Increasing?*, Energy Brief, June 11, 2008, http://tonto.eia.doe.gov/energy_in_brief/natural_gas_production.cfm.

[42] See, for example, Navigant Consulting, Inc., *The State of North American Natural Gas Supply: Review of Groundbreaking Findings,* Presentation to the California Energy Commission, 2009 Integrated Energy Policy Report Workshop, Sacramento, CA, May 14, 2009, http://www.energy.ca.gov/2009_energypolicy/documents/2009-05-14_workshop/presentations/05_Navigant_Pickering_Smead_Natural_Gas_Supply_Assessment.pdf.

[43] See, for example: Elizabeth Bluemink, "Gas Pipeline Project Faces 'Fierce' Competition," *Anchorage Daily News*, September 15, 2009.

[44] Rena Delbridge, "State Sticks by Pipeline Deal," *Alaska Dispatch*, October 26, 2009.

[45] Alaska Dept. of Revenue and Dept. of Natural Resource, *Alaska Pipeline Project Report: Licensed under the Alaska Gasline Inducement Act (AGIA)*, October 31, 2009. p. 21.

[46] Tim Bradner, "Wobbly Market May Crimp Gas Line," *Alaska Journal of Commerce*, February 2, 2009.

[47] A. E. Barinov, "Systemic and Political Factors Affecting Cost Overrun in the World Economy's Large Investment Projects," *Studies on Russian Economic Development*, Vol. 18, No. 6, 2007, pp. 650–658.

[48] Bent Flyvbj erg, Mette K. Skamris Holm And Søren L. Buhl, "What Causes Cost Overrun in Transport Infrastructure Projects?," *Transport Reviews*, Vol. 24, No. 1, January 2004, p. 16.

[49] "Former Canadian Official: State Gas Line Faces Numerous Hurdles," *Associated Press*, April 15, 2008.

[50] Information Insights, Inc., *Economic, Fiscal and Workforce Impacts of Alaska Natural Gas Projects*, Prepared for Alaska Department of Revenue, May 10, 2006, p. 8. Adjusted to 2009 dollars.

[51] Resource Development Council for Alaska, Inc., "Alaska's Oil and Gas Industry," Web page, November 11, 2009, http://www.akrdc.org/issues/oilgas/overview.html.

[52] Ibid.

[53] Information Insights, Inc., May 10, 2006, p. 82.

[54] U.S. Congress, House Committee on Energy and Commerce, Subcommittee on Energy and Air Quality, May 6, 2004, p. 10.

[55] Political Economy Research Institute, How Infrastructure Investments Support the U.S. Economy: Employment, Productivity and Growth, January, 2009, p. 25.

[56] Section 7 of the Natural Gas Act authorizes FERC to issue certificates of "public convenience and necessity" for "the construction or extension of any facilities ... for the transportation in interstate commerce of natural gas" (15 U.S.C. § 717f).

[57] Steven Findlay, Vice President, Denali—The Alaska Gas Pipeline, LLC., *September Monthly Status Report*, FERC Docket PF08-26, Accession No. 20091007-5035, October 7, 2009, p. 2.

[58] TransCanada Alaska Company LLC., *August Monthly Status Report,* FERC Docket PF09-1 1, Accession No. 20090915-5025, September 15, 2009; Tim Bradner, "Gas Pipeline Plans On Schedule But Bumps Lie Ahead," *Alaska Journal of Commerce*, September 18, 2009.

[59] U.S. Energy Information Administration, March 2009, p. 77.

[60] Stefan Milkowski, "Parnell Chamber Talk Focuses on Energy," *Petroleum News*, Vol. 14, No. 40, October 4, 2009.

[61] Federal Energy Regulatory Commission, *Eighth Report to Congress on Progress Made in Licensing and Constructing the Alaska Gas Pipeline*, August 26, 2009, p. 6.

[62] Tony Palmer, Vice President, TransCanada Corp., Testimony before the Alaska House Resources Committee, June 23, 2009.

[63] Bill Walker, All-Alaska Governor, Web page, October 15, 2009, *http://www.billwalkerfor governor.com*.

[64] U.S. Energy Information Administration, April 2009, pp. 42, 44.

[65] U.S. Energy Information Administration, April 2009, pp. 42, 44.

[66] Federal Energy Regulatory Commission, "Existing LNG Terminals," and "Approved North American LNG Import Terminals," Web pages, October 27, 2009, http://www.ferc.gov/industries/lng.asp.

[67] See, for example: John D. Podesta and Timothy E. Wirth, *Natural Gas: A Bridge Fuel for the 21st Century,* Center for American Progress, August 10, 2009, *http://www.americanprogress.org/issues/2009/08/pdf/naturalgasmemo.pdf;* Rena Delbridge, "Federal Energy Commissioner Checks on Alaska Pipeline Projects," *Alaska Dispatch*, September 25, 2009.

[68] For further discussion and analysis of U.S. greenhouse gas policy, see CRS Report RL34513, *Climate Change: Current Issues and Policy Tools*, by Jane A. Leggett.

[69] U.S. Geological Survey, "Assessment of Undiscovered Oil and Gas Resources of the Mackenzie Delta Province, North America, 2004," Fact Sheet 2006-3002, March, 2006.

[70] Thomas R. Berger, Commissioner, *Northern Frontier, Northern Homeland: The Report of the Mackenzie Valley Pipeline Inquiry, Volume One*, Ministry of Supply and Services Canada, April 15, 1977, p. xiii.

[71] Federal Energy Regulatory Commission, *Seventh Report to Congress on Progress Made in Licensing and Constructing the Alaska Gas Pipeline*, February 20, 2009, p. 6.

[72] Claudia Cattaneo, "Little Hope in the Pipeline," *Financial Post*, July 1, 2009.

[73] John Ivison and Carrie Tait, "Pipeline Dream in Peril," *National Post*, October 27, 2009.

[74] Jeffrey Jones, "Imperial Oil CEO 'Dismayed' by Mackenzie Report," *Reuters*, November 3, 2009.

[75] National Energy Board Canada, "National Energy Board Announces Next Steps In Mackenzie Gas Project Hearing," Press release, October 7, 2009.

[76] See, for example: James Irwin, "Alaska Pipeline Advance Could Threaten Mackenzie Valley Pipeline," *Natural Gas Week*, March, 7, 2005.

[77] U.S. Congress, House Committee on Energy and Commerce, Subcommittee on Energy and Air Quality, *Alaska Natural Gas Pipeline Status Report*, 108th Cong., 2nd sess., May 6, 2004, 108-82 (Washington: GPO, 2004), p. 15.

[78] Title I, Division J § 114.

[79] U.S. Department of Energy, "Alaska Natural Gas Pipeline Loan Guarantee," 70 *Federal Register* 30707-30708, May 27, 2005.

[80] Federal Energy Regulatory Commission, August 26, 2009, p. 8.

[81] Stephen Barlas, "Energy Bill Likely to Bless Alaska NG Pipeline," *Pipeline & Gas Journal*, November 2002; "U.S. Loan Guarantees for Alaska Pipeline Worry Mackenzie Line Supporters," *CBC News*, May 27, 2009.

[82] Congressional Budget Office, *S. 14 Energy Policy Act of 2003, As Introduced on April 30, 2003,* Cost Estimate, May 7, 2003, p. 7, http://www.cbo.gov/doc.cfm?index=4206&type=0.

[83] For further background see archived CRS Report RL33629, *BP Alaska North Slope Pipeline Shutdowns: Regulatory Policy Issues*, by Paul W. Parfomak.

[84] National Research Council, *Cumulative Environmental Effects of Oil and Gas Activities on Alaska's North Slope*, Committee on the Cumulative Environmental Effects of Oil and Gas Activities on Alaska's North Slope, National Academies Press, Washington, DC, 2003.

[85] For example, see Jim Springhetti, "State Asks Court to Toss Bradwood Site's Approval," *The Oregonian*, January 26, 2009.

[86] IHS Cambridge Energy Research Associates, *Growth in the Canadian Oil Sands: Finding the New Balance*, Cambridge, MA, 2009, pp. III-1-III-20; Andrew Nikiforuk, *Dirty Oil: How the Tar Sands are Fueling the Global Climate Crisis*, Greenpeace, September 2009.

[87] "Already Hard at Work on Security, Pipelines Told of Terrorist Threat," *Inside FERC*, McGraw-Hill Companies, January 3, 2002.

[88] For additional analysis of pipeline security issues, see CRS Report RL33347, *Pipeline Safety and Security: Federal Programs*, by Paul W. Parfomak.

[89] Elise Stolte, "EnCana Puts Record $1M on Bomber's Head," *Edmonton Journal*, July 3 1, 2009.

[90] D. S. Cloud, "A Former Green Beret's Plot to Make Millions Through Terrorism," *Ottawa Citizen*, December 24, 1999, p. E15.

[91] Y. Rosen, "Alaska Critics Take Potshots at Line Security," *Houston Chronicle*, February 17, 2002.

[92] W. Loy, "Web Post Urges Jihadists to Attack Alaska Pipeline," *Anchorage Daily News*, January 19, 2006.

[93] U.S. Attorney's Office, Middle District of Pennsylvania, "Man Convicted of Attempting to Provide Material Support to Al-Qaeda Sentenced to 30 Years' Imprisonment," Press release, November 6, 2007; A. Lubrano and J. Shiffman, "Pa. Man Accused of Terrorist Plot," *Philadelphia Inquirer*, February 12, 2006, p. A1.

[94] Alyeska Pipeline Service Co., Internet page, Anchorage, AK, November 2, 2009, http://www.alyeska-pipe.com/ about.html.

# CHAPTER SOURCES

The following chapters have been previously published:

Chapter 1 – This is an edited reformatted and augmented version of a United States Geological Survey publication, report 2008-1161, dated 2008.

Chapter 2 – This is an edited reformatted and augmented version of a Sandia National Laboratories publication, report SAND2009-1044, dated February 2009.

Chapter 3 – This is an edited reformatted and augmented version of a Congressional Research Service publication, report R40963, dated November 30, 2009.

# INDEX

## C

## D



## E

## I



## J



## K



## L



## M

## N



## O



## P

## Q



## R



## S

## T

# THE BEST OF INDIAN PHYSICS

## CONTEMPORARY FUNDAMENTAL PHYSICS

***Valerie V. Dvoeglazov***
***Series Editor***

**Symmetry and Electromagnetic Interaction of Fields with Multi-Spin**
*S. I. Kruglov*
2000. ISBN 1-56072-880-9

**Some Contemporary Problems of Condensed Matter Physics**
*Stoyan J. Vlaev and L. M. Gaggero Sager*
2000. ISBN 1-56072-889-2

**The Photon**
*Trilochan Pradhan*
2001. ISBN 1-56072-928-7

**The Problem of Electron and Superluminal Signals**
*V.P. Oleinik*
2001. ISBN 1-56072-938-4

**Chaotic Universe: From the Planck to the Hubble Scale**
*B. G. Sidharth*
2001. ISBN 1-56072-977-5

**The Best of Indian Physics**
*B. G. Sidharth*
2005. ISBN 1-59454-115-9

**Process Physics: From Information Theory to Quantum Space and Matter**
*Reginal T. Cahill*
2005. ISBN 1-59454-300-3

# THE BEST OF INDIAN PHYSICS

B. G. SIDHARTH
EDITOR

Nova Science Publishers, Inc.
*New York*

**LIBRARY OF CONGRESS CATALOGING-IN-PUBLICATION DATA**

The Best of Indian Physics / B. G. Sidharth, editor.
p. cm.
Includes index.
ISBN 1-59454-115-9 (hardcover)
1. The Best of Indian Physics. B. G. Sidharth.
QC173.7.T74 2004
530.14'072--dc22 2004024758

*Published by Nova Science Publishers, Inc. ✣ New York*

# CONTENTS

# PREFACE

India has had a very long tradition of path-breaking excellence in science and technology including mathematics, physics and astronomy. Even though the British rulers of India had no enthusiasm for Indians learning and doing science, the rising tide of nationalistic fervor in the first half of the twentieth century gave rise to, what has been called the golden age of Indian physics. There were some physicists, amongst others, who could be ranked with the best in the world. If we are to speak of the Indian Physics, then four names from that period immediately come up: Jadadeesh Chandra Bose who, before Marconi, actually demonstrated the propagation of radio waves. Some years later Sir C.V. Raman received the Nobel Prize for his celebrated Raman Effect. Then there were Meghnad Saha and Satyen Bose who made seminal contributions to physics. This new and important book offers a window into the backgrounds and achievements of these remarkable scientists.

In: The Best of Indian Physics
Editor: B. G. Sidharth, pp. 1-6
ISBN 1-59454-115-9


*Chapter 1*

# THE LEGACY

***B. G. Sidharth***
International Institute for Applicable
Mathematics and Information Sciences, India

The eleventh century historian of science Syed Al Andalusi wrote in his "Quitab Taba Kat al Uman" (Categories of Nations), that of all the countries in the world India was scientifically the most advanced. This was not surprising because already the Indian mathematician - astronomers were using principles of Differential Calculus in their calculations of planetary positions and motions. These included scholars like Munjala, Sripati and Bhaskaracharya. All this was some six hundred years before Newton and Liebniz independently invented Differential Calculus in Europe.

Of course India's pioneering contributions in mathematics are well known. Thus the decimal system, or the scale of ten notation with a symbol for zero was in vogue for thousands of years, before it reached Europe around the thirteenth century via the Arabs. A thousand years before Newton and even earlier, mathematicians like Bramhagupta and Aryabhatta were using quite advanced mathematical techniques including Trigonometry and interpolation via second order difference tables in their astronomical calculations. Geometry too had been very well developed, as it was needed in the construction of religious altars. For example, the so called Pythagoras theorem was already known in India

Indian mathematics reached the Arabs in the eighth century itself through scholars like Kanka of Ujjain, who was invited to Baghdad by Khalif Al Mansur. Here he translated Brahmagupta's work into Arabic with the help of an Arabic scholar, Al Fazari. The translation, Hind Sind is still available. But within a few centuries, Indian mathematics and science was in decline, due to foreign invasions. Nevertheless, some studies continued in the South. Notable are names like Madhava and Nilakantha. The latter had worked out Gregory's series for the inverse tangent, a few centuries before Gregory.

What is not so well known is that India had a very long tradition, not just in mathematics, but also in science and technology including astronomy. Indeed, mathematics does not evolve in isolation. It grows, at least to some extent, for fulfilling societal needs. The problem has been that there are few surviving specialist texts, though here and there remnants of science and technology appear in traditional arts and literature: allusions to advanced concepts in topics like astronomy and physics are found scattered in the vast body of ancient literature, which give glimpses of a hoary, flourishing tradition.

For instance there are many interesting astronomical references in the Rig Veda, which is more than four thousand years old, and could be as old as ten thousand years. In any case it is the oldest extant Indo-European literature. There is quite a bit of calendric astronomy in the Rig Veda, because the calendar was central to the newly born agrarian civilizations. But this apart, there are amazing statements like the earth being a bowl, that is rotund, the sun being a star of the daytime and thoughts on why the sun doesn't crash down, even though it is not supported. (It may be mentioned that much later, the Greeks also asked such a question, but they invented crystalline transparent spheres to support the sun and other objects). Similarly there is a clear recognition of the fact that the stars, though appearing to be at the same distance, are in fact at differing distances. (For the Greeks, in a much later epoch, all the stars were en masse attached to a single sphere).

Consider the following. In an amazing dialogue in the Mahabharata, Bhrigu says:

> "the sky thou seest above is infinite.... its limits cannot be ascertained. There where the rays of the sun and the moon cannot reach are self-luminous objects which possess splendour like that of the sun. Even these latter cannot access the limits of the firmament in consequence of the inaccessibility and infinity of these limits...This place is full of many blazing and self-luminous worlds...."

A problem, in a physical context, that troubled early thinkers was whether trees and plants are living creatures, as they do not seem to have any of the attributes of animals, for example they are dense, in the sense that there are no spaces and liquids inside them, they are not mobile, they don't seem to eat food and so on. In the Mahabharata, Bharadwaja asks of Bhrigu, just quoted:

> "Trees do not appear to have any heat. They do not seem to have any motion. They are again made up of dense particles... Trees do

not hear. They do not see. They are not capable of the perception of scent or taste. They have not also the perception of touch. How can they then be regarded as "living creatures"...It seems to me that in consequence of the absence of any liquid material in them, of any heat.... any empty space..."

To which Bhrigu replies:

"Without doubt though possessed of density, trees have space within them. The putting forth of flowers and fruits is always taking place in them. They have heat (energy) within them in consequence of which leaf, bark, fruit and flower are seen to droop. They sicken and dry up... A creeper winds round a tree and goes about all its sides... Then again trees recover vigour and put forth flowers... They drink water by their roots. They catch diseases of diverse kinds. Those diseases again are cured by different operations... As one can suck up water through a bent lotus stalk, trees also drink through their roots.... They are susceptible of pleasure and pain and grow when cut or lopped off... From these circumstances I see that trees have life."

(Incidentally it is interesting to note that there is an ancient system of Indian medicine for curing plant diseases.)

Finally it may be mentioned that in ancient India there was knowledge of microorganisms. In a passage of the Mahabharata it is mentioned that one kills life even while eating grains like rice, because these microorganisms are contained in the rice; and that when one simply breathes or walks he interacts with these microorganisms and may even kill them.

There are references to the water cycle here and there, for instance in the Aditya Hrudayam of the Ramayana. There are clear references to the sun absorbing water from the earth and raining it back to the earth. So also books like the Vishnu Purana indicate knowledge of the tides. For instance,

"In all the oceans the water remains at all times the same in quantity, and never increases or diminishes; but like the water in a caldron, which, in consequence to its combination with heat, expands, so the water of the ocean swells with the increase of the moon. The waters, although really neither more nor less, dilate or contract as the moon increases or wanes..."

In fact there is still a lingering tradition in India that makes it inauspicious to undertake travel on new moon and full moon days, a tradition which can be directly traced to the knowledge of the tides, remembering that in ancient times lengthy travel involved the crossing of rivers and seas.

In many ancient texts there are references to Atomism. For instance, to quote the Vishnu Purana again,

> "How can reality be predicated of that which is subject to change, and reassumes no more its original character? Earth is fabricated into a jar; the jar is divided into two halves; the halves are broken to pieces; the pieces become dust; the dust becomes atoms..."

In fact at least about seven centuries before Christ, Atomism and Vibrations were known through the works of scholars like Kanada. It would be interesting to survey a few thoughts on what we today call Physics, from Kanada [2]:

1. Action is opposed by its effect, reminiscent of Newton's third law.

2. Colour is the combined effect of many colours.

3. Air is made up of many substances (and is not one homogenous substance).

4. The first action of the arrow is from impulse; the next is from resultant energy produced by that action... in the absence of propulsive energy generated by action, falling results from gravity.

5. The sun's rays cause the ascent of water, through conjunction with air (atmosphere).

6. Condensation and dissolution of water are due to heat.

7. Thunder clap results from conjunction with and disjunction from water of the cloud.

8. Largeness or magnitude is produced from multiplicity.... the contrary of this is the atomic.

9. However small and large are relative.

10. Atoms can be conjoint.

11. Bodies are formed by, "ultimate atoms, inconstant in direction and position" (i.e. vibrating).

It may be mentioned that the last point, that is that atoms are vibrating (and are not static) is unique to Indian thought. Equally amazing are the time intervals, which are defined, in ancient Indian texts. Starting from a period like 8640 million years, which is defined to be the life of the universe, tiny periods up to less than a five hundredth of a second have been defined.

Sounds and vibrations had been studied in the context of music, for instance the seven primary musical notes, and also in a more general context. (In fact the Nobel Prize winning Indian Physicist, Sir C.V. Raman devoted much of his studies to the acoustics of Indian musical instruments). For instance in the Bruhadaranyaka Upanishad, it is stated,

> "When a drum is beaten, one cannot distinguish its various particular notes, but they are included in the general note of the drum or in the general sound produced by different kinds of strokes"

and similar verses. Clearly the principle of superposition of notes had been well established.

In the Mahabharata it is mentioned that stars appear to be so small, only because of their great distance while the Markandeya Purana speaks of the equatorial bulge of the earth [4,5,6]. Similarly there are references to the so called fixed pole star moving away, that is to precession of the equinoxes. The Vishnusahasranama, a part of the Mahabharata clearly describes the sun as supporting the earth. There are any number of such amazing references strewn all across the ancient texts of India.

Another field of technology where India outshone other countries was Metallurgy, and this became to be recognized in recent times. For instance some two thousand years ago zinc smelting on a large scale had already been perfected. The American Society of Metals has recognized this as an "International Historical Landmark for Metallurgy". In fact it is said that when Alexander the Greek reached India, he wanted as a present, some Indian Steel. A monument of India's ancient metallurgical prowess is the amazing iron pillar in Delhi, which is at least about sixteen hundred years old, and stands to this day in the open air, without any rust, a symbol of a technology that has with other such technologies been lost over the centuries. In more recent times, the so called Damascus steel for swords made in Damascus was in fact imported from India. The English word wootz can be traced back to a local Indian word for steel, "ukku". There are crucibles and tangible traces of its manufacture, available in India even today. Similarly the more than a thousand year old bronze technology for casting of icons, a continuing tradition to this day in South India is another marvel. It is said that Bessemer the German Metallurgist openly gave credit to rural Indian technology, for the process that goes after his name. Indeed this is true in many other fields, for instance textiles. Or, as another example vaccination which Edward Jenner came to know of from India.

A curious fact, which is not well known is that, the trigger for modern rocketry came from India. India had a long history of rocketry and fireworks, a tradition that continues to this day. In the battles against the British more than two hundred years ago, the Indian armies used these rockets freely. Even today there are replicas of these earliest rockets in British Museums. For instance the Science Museum in London exhibits one of the earliest rockets encountered by the British. This had a range of one kilometer and was used by the Marathas. The British shipped these rockets back to England, studied them and even made their own rockets. But then the knowledge spread to other parts of Europe. All this culminated in the V2 German rockets of the Second World War vintage, which again were the trigger to present day rockets.

In: The Best of Indian Physics
Editor: B. G. Sidharth, pp. 7-18
ISBN 1-59454-115-9


*Chapter 2*

# CALCUTTA TRIO

***B. G. Sidharth***
International Institute for Applicable
Mathematics and Information Sciences, India

## 1 C.V. Raman

The Assistant Accountant General of the Government of India was riding to his office in a Calcutta (now Kolkata) tram. Suddenly he spotted a board as the tram crossed Bow Bazaar Street. The legend read, "Indian Association for the Cultivation of Science." Promptly the officer got off at the stop, walked up to the place and knocked on the door. Little did he realize that he was knocking on the doors of a Nobel Prize.

For, the Assistant Accountant General was Chandrasekhara Venkataraman, known popularly as C.V. Raman. Raman was born near Tiruchirapalli in South India on the 8th of November 1888. When he was four years old, his father R. Chandrasekhara Iyer joined the A.V.N. College in the South Eastern city of Visakhapatnam, to teach Physical Sciences. The young Raman was a voracious reader. One of the books that left a lasting impression on him was "Sensation of Tone", written by the well known German Physicist Helmholtz.

When he was thirteen, Raman left Visakhapatnam to study at the Presidency College, Madras on a scholarship. Though the youngest student there, he also shone as the best, particularly in English and Physics. On graduation, Raman was persuaded to attempt the Indian Civil Services examination of those days. For this very competitive test he had to go to England, but first had to qualify a medical examination. Raman failed the test. This was a bitter disappointment at that time, but as Raman recounted later about the Civil Surgeon who had disqualified him, "I shall ever be grateful to this man."

Raman took up Post Graduate studies in Physics. Already his brilliance was being recognized. In Raman's words,

> "Prof. Jones (of Physics) believed in letting those who were capable of looking after themselves to do so, with the result that ... I enjoyed a measure of academic freedom that seems almost incredible..."

In fact during the two year course Raman attended only one lecture!

As a result Raman could perform his own experiments with light. A short paper of his was even published in the well known Philosophical Magazine. His laboratory adventures enabled him to write another paper on measurement of surface tension of liquids.

After completing his degree in early 1907, with the maximum distinctions available, he appeared for the very competitive Financial Civil Services Examination of those days. Again he was a topper.

And that is how the teen aged Raman, with his newly married wife joined as the Assistant Accountant General in Calcutta in 1907 itself.

At the Indian Association for the Cultivation of Science in Bow Bazaar Street, Raman was welcomed with open arms, because he was the first person in a long time to evince interest in this institution, which had been set up with patriotic fervor in 1876 by the late Mahendralal Sircar, so that the native Indians could do science as well. Soon Raman shifted to a house close by. His day would begin at 5.30 a.m., when he would go to the Association, to return after some four hours for a quick bath and meal before catching a horse drawn taxi, in order to be in time for his office. Raman would return to the Association directly from work at 5.00 p.m. to work for another four to five hours. He devoted Sundays fully for work at the Association.

During the next ten years, Raman's research findings appeared in Nature, the Philosophical Magazine and the Physical Review. The work dealt mostly with Vibrations and Acoustics. Already he had earned a reputation, not just in Calcutta, but also in Europe and USA. He had studied various Indian musical instruments and their acoustical and vibrational patterns. But ironically, Raman through all these years was a full time civil servant, a conscientious Assistant Accountant General who earned a very good name for himself in his job. The British bosses believed that Raman would rise to the highest available position in the Finance Department.

Around this time, somewhere around 1916, Raman was offered the Palit professorship in Physics at the Calcutta University. The catch was that academics being what they were, the salary was about half what he was drawing in government service. Raman forsook wealth and position for the cause of his love, Physics. He became the Palit Professor of Physics in Calcutta University in 1917. This was a research position but Raman also took courses. In the words of Ramdas a student who later distinguished himself,

> "Prof. Raman took `Electricity and Magnetism' in the year 1920-21 and `Physical Optics' in 1921-22 ... students felt that they were indeed listening to a type of inspired teaching to which was brought all the original flavour and excitement of the great giants of the past."

By now Raman was attracting research scholars from all over India. The list included names like K.R. Ramanathan, K.S. Krishnan and Suri Bhagawantham, all of whom distinguished themselves in later years.

In 1919, Raman became the Honorary Secretary of the Indian Association for the Cultivation of Science. He had transformed a forsaken place into the hub of experimental activity. Around this time Raman started seriously looking into optics. In 1921, Raman went abroad for the first time, after obstinately refusing to do so earlier, even when legally required to do so. He was representing Calcutta University at the University Congress at Oxford.

In September 1921, the S.S. Narkunda left Southampton on its two week voyage to Bombay with Raman on board.

While on the cruise, Raman performed his optical experiments, fascinated as he was by the different hues of the sea. At Aden, while other passengers went to explore the place, Raman went to the post office to deliver a paper to Nature. This paper investigated the colours of the sea. He investigated the question, why does the sea look blue?

This question was thought to have been answered. Lord Rayleigh had already explained the blue of the sky. This was because the blue component of sunlight is scattered a lot more than the other colours. And, the sea would be blue, because it was merely reflecting the blue sky. But Raman had other ideas. Back in Calcutta Raman plunged into further experiments and worked on this topic. He concluded that the blue of the sea was not because of the reflection of the sky as was universally believed, but rather it was because of the scattering of light by the water molecules, as in the case of the sky.

By 1922 Raman came out with a paper, "The molecular diffraction of light" wherein for the first time he considered the possibility of the scattering of light by molecules, light now being treated as corpuscles or photons. Ironically, this idea had been put forward by Max Planck almost two decades earlier, and Einstein had even used it in his explanation of the photo electric effect. This was also at the heart of the explanation for spectral lines given by Niels Bohr. But when it came to scattering, light was considered in its original classical context of Maxwell's Electromagnetism, that is, as in Rayleigh's scattering.

The next year Compton discovered the effect that goes by his name. In the Compton effect, X rays behave like corpuscles or photons. Raman's experiments were now investigating the possibility of visible light waves also having a similar behaviour. He did not have the sophisticated light sources that were being used in laboratories in other parts of the world, and made up for it by using sunlight itself with filters wherever necessary. Very early on, in 1923 itself, in one of Raman's experiments, his student K.R. Ramanathan observed for the first time that light scattered from a liquid showed a change in colour. Though this effect was not very pronounced, repeated experiments confirmed the finding. About the only explanation possible was that it was some sort of a fluorescence in which light of a certain frequency or colour is absorbed by certain materials and re-emitted. Raman and his Group called it Feeble Fluorescence. In 1925, another student K.S. Krishnan in fact observed this feeble fluorescence in fifty other liquids, which made the phenomenon all the more puzzling. Could it be fluorescence at all? If not, what was it? The point is that flourescence is caused only by certain material, and that too by the impurities present therein. The Feeble Fluorescence effect persisted even as these liquids were being substituted by purer and purer samples. Around this time, another student, S. Venkateshwaran found that when sunlight was passed through highly purified glycerine the scattered light had a brilliant green colour instead of the expected blue.

Raman was now an inspired person. He got these experiments repeated by his team. By early February 1928 the results of the experiments were clear: Several pure liquids displayed a fluorescence like effect. Very interestingly the scattered light was strongly polarised. Something which was not characteristic of fluorescence. On the 16th of February Raman communicated a paper to Nature entitled "A New Type of Secondary Radiation". However the experiments continued. On the 28th of February, 1928, in the words of K.S. Krishnan,

> "...we proceeded to examine the influence of the incident wavelength on the phenomenon. Used the usual blue-violet filter coupled with uranium glass, the range of wavelengths transmitted by the combination being much narrower than that transmitted by the blue-violet filter alone. On examining the track with a direct vision spectroscope, we found to our great surprise that the modified scattering was separated from the scattering corresponding to the incident light by a dark region."

It may be mentioned that K.S. Krishnan subsequently distinguished himself as a great teacher, not only in Calcutta, but also Dacca, Allahabad and Delhi. He joined Raman in Calcutta in 1923. Later, in 1928 he moved on to Dacca University to the Department of Physics which already boasted of Satyendranath Bose. He returned to the Indian Association for the Cultivation of Sciences, Calcutta, as the Mahendralal Professor of Physics in 1933, and later joined the Allahabad University as Professor and Head of the Physics Department. He became the Director of the National Physical Laboratory, in Delhi in 1947, where he remained till the very end. Apart from his work on Optics, starting with his collaboration with Sir C.V. Raman, he also worked on the behaviour and properties of crystals, lattice dynamics and thermionic emission, with great distinction.

Raman immediately realised that what they had been observing was a new type of scattering, with a change of frequency or colour of the light, something that could be explained only on the basis of the Quantum Mechanical photon, and something that was similar to the Compton effect.

This was the Raman Effect, and the 28th of February is celebrated as the National Science Day in India.

Raman announced the discovery at a Press Conference the very next day, and recognitions started pouring in quick and fast. These culminated in the Nobel Prize in 1930. A local news agency informed Raman's student, Bhagawantham about it, who promptly conveyed the news to Raman.

Sir C.V. Raman received the Nobel Prize on the 10th of December 1930 in Stockholm. An American diplomat present at the function wrote:

> "Of all the prize winners, the day was easily carried, however, by Sir Venkata Raman the Indian Prize winner, who, upon returning to his seat on the platform after receiving the prize from the hands of the King, was visibly moved by his emotion and sat with tears streaming down his face...

At the banquet that evening, Sir Venkata Raman's speech was a masterpiece of eloquence which called forth tremendous applause from a banquet-weary gathering not noted for its responsiveness. Less appreciative was perhaps the British Ambassador, who sat one place removed from me, who was forced to listen with equanimity to Sir Venkata Raman's reference, brief though it was, in passing only, to the congratulatory telegram which he received "from his dearest friend now in jail (Mahatma Gandhi)..."

Later Raman reminisced that when he saw the Union Jack on his seat he was overcome by the emotion that not his beloved country India, but rather a colony of Britain was represented.

After more than almost seventy years after his discovery, a huge body of literature (including an International Journal on Raman Spectroscopy) continues to pour in. In retrospect, the secret behind the Raman effect was the following: Even though the Quantum nature of energy and electromagnetic radiation was known at that time, it was still the wave nature of light and the classical Rayleigh's scattering that was being considered. Raman realized that a la the Compton effect, it was the Quantum nature of light that had to be invoked to explain what could otherwise pass off as anomalous Feeble Fluorescence.

Around 1933 the Indian Institute of Science in Bangalore was looking for a Director to replace the then retiring head, Dr. Martin Forster. On the recommendation of Ernest Rutherford, he was appointed the Director. For the first time an Indian occupied that position.

Raman had a rough innings. He tried to breathe new life and dynamism into a soporophic Institute, which didn't go very well. The last straw was that Raman invited the famous and eminent Physicist Max Bohm to the Institute. Prof. Aston and others at the Institute strongly opposed the coming of Max Bohm, who in their eyes was a German non-entity. So it was that through these machinations Max Bohm returned to England and Raman had to step down from the post of Director in 1938, but continued as a professor till 1948.

Meanwhile Raman founded the Indian Academy of Sciences in Bangalore, which today has blossomed into an elite club of Indian scientists. All through Raman continued his researches in Optics and his other areas of interest . In particular work with his student at the Institute, Nagrendranath threw up the well known Raman Nath theory. This theory was the inspiration for the work of Berry, of the Berry phase fame, in England many years thereafter.

After retirement in 1948, Raman established a Research Institute, now called the Raman Research Institute in Bangalore. He continued to be busy with optical experiments. He also studied and collected crystals including diamonds. His pupil Dr. S. Bhagawantham, who later rose to very high levels of science administration in the country also distinguished himself in Crystallography. Unfortunately the last years of Raman's life were mired in controversy which was triggered off in part by his own strong, frank, no nonsense views on Indian science. Thus he was not happy with the proliferation of elite scientific laboratories with sprawling campuses and, by the standards of those days, quite luxurious buildings. He thought all this was a waste and called it the Nehru-Bhatnagar Effect-Dr. S.S. Bhatnagar being the Head of India's Council of Scientific and Industrial Research in the early years of independence. Uppermost in his mind was that India had been left a very impoverished country and could not afford too many luxuries. To his very last Raman was a great nationalist and a great scientist.

# 2 Meghnad Saha

Meghnad Saha, another doyen of Physics, both at the International and the Indian levels, is best known for his work on the ionization potential. This laid the foundation for very many studies in Astrophysics, dealing with stars. He was born on the 6th of October 1893. His father Jagannath was a petty shopkeeper near Dhaka, today the capital of Bangladesh. In his early school days, Meghnad shone to such an extent that his teachers and even some family members were keen to see him take admission to an English medium school, which was some ten kilometers away in another village. So, much against his father's will, and helped by the free boarding and lodging of a sponsor, Ananth Kumar Das, Saha continued his studies.

Later he joined the Government Collegiate School in Dacca (now Dhaka), on a free studentship. Unfortunately Saha lost the scholarship because he joined a boycott movement against the British, which arose from the partition of Bengal in the early years of the twentieth century. After a few years of struggle, finally Saha topped the list of candidates at the collegiate test and so joined the Intermediate Dhaka College for the two-year Intermediate course.

Upon its completion the young Meghnad joined Presidency College, Calcutta, for the B.Sc Degree helped by a scholarship. One of his distinguished classmates was S.N. Bose. A distinguished teacher was Jagdeesh Chandra Bose. On completion of the degree, Meghnad who was a brilliant student, wanted to appear for the Financial Civil Services Examination, so that he could support his hard up family. Unfortunately, or fortunately, as it turns out in many cases, he was debarred from the examination because he was involved with the freedom movement. At this stage Meghnad had to really struggle for survival, living on private tuitions. Luckily Sir Ashutosh Mookerjee appointed both Saha and Bose in the Calcutta University around 1916. Soon they were with Raman who was by then the Palit Professor. Thereafter Saha was completely immersed in teaching and research.

Starting from 1917, Saha published a number of papers on Radiation and other topics. In 1918 he was awarded the D.Sc of the Calcutta University and soon got married shortly. Thereafter a number of papers on ionization and the spectra of stars tumbled out, triggered off by a paper of Eggert, then a student of the famous German Physicist Nernst. The story really went back to a study of the Solar Chromosphere. The British Physicist, Sir Norman Lockyer had noted that the flash spectrum of the Chromosphere of the sun was enhanced due to the prevailing high temperature.

Eggert tried to apply the principles of heat to investigate this phenomenon, though unsuccessfully. Saha realized that the enhancement of the spectral lines would be due to ionization, rather than heat or temperature. Where Eggert had failed to calculate the heat of the reaction, Saha could do so by utilizing the ionization potential and thus deduced a relation between the fraction of the ionized atoms, the heat of reaction and the temperature and pressure.

Saha wrote an essay, "On the Harvard Classification of Stellar Spectra" which won him the Premchand Roychand and Guruprasanna Ghosh Fellowships in 1919 itself. This enabled him to travel to Europe immediately. Saha survived in London thanks to the timely help from A. Fowler of the Imperial College. With advise from Fowler, Saha rewrote his prize winning essay. This was communicated to the Royal Society by Prof. Fowler and accepted for publication.

The paper immediately was appreciated all around. Saha generously acknowledged Fowler's help. As he reminisced,

> "I took about four months in rewriting this paper, and all the time I had the advantage of Professor Fowler's criticism, and access to his unrivaled stock of knowledge of spectroscopy and astrophysics. Though the main ideas and working of the paper remained unchanged, the substance matter was greatly improved on account of Fowler's kindness in placing at my disposal fresh data, and offering criticism whenever I went a little astray out of mere enthusiasm."

In 1920 itself, Saha went to Germany to the Laboratory of Nernst and performed experiments for a whole year. Curiously enough it was here that the well known Physicist Sommerfeld introduced Saha to the Nobel Prize winning Indian poet Rabindranath Tagore.

Later Saha went to Switzerland before returning to England. By this time Saha knew that his work would be of utmost importance in understanding not just the sun but stellar spectra in general. He also realized that much of spectroscopic data was available in USA, thanks to Observatories like Mount Wilson. Though Saha tried to visit America for this purpose, he did not get any support for doing this, despite the fact that American Astrophysicists like H.N. Russel fully realized the importance of his work. Russel in fact wrote,

> "I believe that within a few years we may utilize knowledge of ionization potentials, and so on, to obtain numerical determination of stellar temperatures from spectroscopic data."

Thereafter Saha returned to Calcutta to become the Khaira professor. Unfortunately Saha could not make much progress here with his experiments due to lack of funds and so, joined the Allahabad University in 1923, thanks to appreciation of his work by Eddington, Einstein and others. Even here the going was tough and Saha could not build up a laboratory to his expectations. In 1926 he was elected a Fellow of the Royal Society. This helped him to obtain a meager grant for his laboratory. The fact is that Saha had to struggle against odds and was crippled by the lack of adequate facilities.

In retrospect, Saha's greatest contribution, the ionization formula, had brought in two factors, which other researchers had failed to introduce. Starting from Lockyer himself, it was felt that the enhanced spectrum was due to heat alone, while pressure was ignored. Next the contribution of ions to the enhanced spectrum was totally ignored. These were two key lacunae, which Saha could recognize.

Swimming against the tide was the dominant feature of his life. Saha now turned to the problems of the ionosphere and radio waves. In a short paper entitled "A Stratosphere Solar Observatory", he put forward the idea of the observation of the solar spectrum using high altitude balloons.

He went on to get the U.P. Academy of Sciences, later the National Academy of Sciences, established in 1932. The idea was to harness science and technology and funnel it to the masses through the universities. For instance, in 1938, he had a symposium organized by the Academy on the theme, "Power Supply."

In 1935 he got the National Institute of Sciences established, originally in Calcutta, but later shifted to New Delhi, again at his instance. Even later it came to be known as the Indian National Science Academy. There is no doubt that Saha was driven by an intensely nationalistic spirit, quite like Raman. For example, that year he lamented,

> "In poverty, ignorance and disease, India of today can only be classed with China and Abyssinia [Ethiopia], countries which are still steeped in medievalism..."

Meanwhile Saha had ventured abroad on brief and not so brief trips, which also impressed upon him the growing importance of Nuclear Physics, an impression that he carried with him till the very end. By this time the Palit Professorship at the Calcutta University had fallen vacant. Firstly Sir C.V. Raman had relinquished his charge in 1934 when he shifted his base to Bangalore and secondly the new incumbent, Prof. D.M. Bose vacated the post to take over as the Director of the Bose Institute in Calcutta. Saha took over the Palit professorship and continued to hold the appointment till his retirement. One of the first things that Saha did was to introduce Nuclear Physics in the university. Using his personal relationship with Jawaharlal Nehru (later India's first Prime Minister), he could obtain a grant towards setting up a Cyclotron at Calcutta University. Unfortunately this proved to be a much too protracted project. However he was successful in establishing an Institute for Nuclear Physics, which got off the ground in 1948.

Saha retired in 1952 from the Palit Professorship and also from the Directorship of the Institute of Nuclear Physics, later named after him.

This was the ideal opportunity to take on the reigns of the flagging Indian Association for the Cultivation of Science. Due to shortage of space Saha got the Association shifted from its original Bow Bazaar location, where in fact Raman made the Nobel Prize winning discovery, to a new and sprawling campus, in a different part of the city. Saha continued as Director of IACS till the very end, in 1956.

The later years of Saha's life were devoted more to pedagogic and organizational activities rather than research per se. He was the member of several national scientific bodies and institutions. These included a programme to reform and reconcile various Indian calendars and evolve a National calendar. Though he was the pioneer of Nuclear Physics in India, he opposed the formation of the Atomic Energy Commission in 1948, on the grounds that first a manpower base and industrial infrastructure needed to be built. This did not go well with the Governments' "do it fast" attitude.

Of humble origins and himself a victim of colonial rule in one way or another, Saha was acutely aware of the requirements of social justice. It is a bit sad that his work on the ionization potential, though path breaking at that time, did not find a prominent place in later developments or in contemporary literature.

## 3 S.N. Bose

The third of the Calcutta trio was Satyendranath Bose who was born on New Year's day in 1894. After graduation, Bose got his Master's degree in Mixed Mathematics (later bifurcated into two departments, the Department of Pure Mathematics and the Department of Applied Mathematics) of the Calcutta University in 1915. Under the patronage of Sir Ashutosh Mukherjee, he joined the Physics Department in a few years. Here Meghnad Saha was his colleague and co-worker. In fact they produced a joint paper on the equation of state which was published in the Philosophical Magazine in 1918. Even today the equation is referred to as the Saha-Bose Equation. Two other papers followed which were published in the Bulletin of the Calcutta Mathematical Society while in 1920, yet another paper of Bose on Rydberg's Law was published in the Philosophical Magazine.

In 1921 he took up a higher position as a Reader in the University of Dacca in those days in the East Bengal region of India. Soon he rose to the rank of Professor and Dean of the Faculty of Sciences. In 1924 Bose wrote a paper on Planck's Radiation Law and sent it to Albert Einstein who was immediately interested and excited. In fact Einstein wrote,

> "Bose's derivation of Planck's formula signifies in my opinion, an important step forward. The method used here gives also the Quantum Theory of an ideal gas as I shall show elsewhere."

The paper was published in Zeitschroft fur Physik and laid the foundation for what is today widely known as Bose Statistics or Bose-Einstein Statistics.

The essential point and novelty of Bose's idea was that in a Quantum Mechanical treatment, particles with spin zero or an integral spin, this including non ionised atoms would tend to cluster together very closely in a way that cannot be understood from a classical point of view. This Bose-Einstein condensation was experimentally realised seventy years later!

This paper was followed by another paper which also appeared in the same journal in 1924. Unfortunately, the translator, none other than Albert Einstein himself, this time remarked in a brief note that he could not agree with some points of the paper. This negative comment unjustifiably put Bose's work on thermodynamic equilibrium in radiation in the presence of matter in the cold. His interest shifted to other problems including the total reflection of electromagnetic waves in the ionosphere, in 1938, while in 1939 he worked on the Lorentz group, and later, on an integral equation formulation of the Hydrogen atom problem and so on.

In 1945 he returned to Calcutta University as the Khaira Professor of Physics. He was elected the President of the National Institute of Sciences of India in 1949. He retired from Calcutta University and joined the famous Vishwa Bharati University, founded by Rabindranath Tagore as its Vice Chancellor. In 1958 he was elected a Fellow of the Royal Society (of England).

Bose kept up his interest in different areas ranging from crystal structure and X-ray spectroscopy to the theory of General Relativity. In fact in the 1950s it was believed, though erroneously, as it turned out later, that Professor Bose had managed to solve the equations of the unified field theory. For instance The Hindu newspaper of June 12, 1953 carried an item:

> S.N. Bose Findings on Unified Field
>
> Professor S.N. Bose, Dean of the Faculty of Sciences in Calcutta University, had solved equations with regard to the General Theory

> of Relativity, till then thought impossible to work on. He deduced some fundamental identities of the Unified Field Theory (General Theory of Relativity) in a general manner without the restrictions introduced by previous workers. Bose's findings were expected to give the Unified Field Theory a fillip and help realize Albert Einstein's expectation that "all results in the Theory of Relativity should follow from rigorous solutions." Prof. Bose had formulated his theory by utilising, his new findings, and verified the new formulation, by showing, that the usual equations in the electromagnetic field could be obtained by his method. Papers containing his findings were to be published in America, France and Italy. The 50-year-old Indian scientist enjoyed an International reputation, having made valuable contributions on Quantum Statistics appreciated by Einstein himself, and the Bose-Einstein Theory was well known. A group of fundamental particles had been named after Bose as "Bosons" by Professor Dirac, Nobel Laureate of Cambridge University.

Meantime in 1948 Professor Bose had founded the Bangaya Bigyan Parishad to popularize science in the mother tongue, a very effective tool in a country where most people do not know English well. He was associated with this forum till his very end.

Though Professor Bose's contributions have been relatively few in number-he had just some 25 publications to his credit during his entire career-they were very fruitful. It is a tribute to his genius that the experimental work on Bose-Einstein condensation has just started in the recent years.

In: The Best of Indian Physics
Editor: B. G. Sidharth, pp. 19-23
ISBN 1-59454-115-9


*Chapter 3*

# SUBRAMANIAN CHANDRASEKHAR

*B. G. Sidharth*
International Institute for Applicable
Mathematics and Information Sciences, India

On his 73rd birthday, Dr. S. Chandrasekhar received a truly fairy-tale birthday present: The 200,000 US Dollar Nobel Prize for Physics, for 1983, which he shared with Prof. William Fowler. According to the citation, he was being honoured for his theoretical studies of the physical processes of the importance of the structure and evolution of the stars. This was work done in the mid thirties!

While accepting the highest scientific award, Dr. Chandrasekhar recalled the memorable lines of Rabindranath Tagore which encapsulate India's age old thinking,

> "where the mind is without fear and the head is held high, where the knowledge is free, where words come out from the depth of truth... into that haven of freedom, let me awake."

Subrahmanyan Chandrasekhar a nephew of Nobel Laureate, Sir C.V. Raman, was an outstanding student of Physics Honours at Presidency College, Madras. His brilliance earned him Sir Arthur S. Eddington's book, "The Internal Constitution of Stars" as a prize. Soon thereafter, Chandrasekhar won a scholarship which took him to Cambridge.

On the steamer, Chandrasekhar whiled the lonely hours figuring out how collapsing stars ended up. He presented his conclusions later, on the 11th January 1935, at a meeting of the Royal Astronomical Society.

He had carefully discussed the topic in advance with the then doyen of Astronomy, Sir Arthur Eddington, who had even arranged for extra speaking time to be given to Chandrasekhar. But Eddington had been reticent about his own comments, saying cryptically, "That is a surprise for you."

Chandrasekhar spoke about the mysterious `white dwarf' stars which had been discovered in the 1860s itself. But they could be understood only decades later. Every star, under its own weight, tends to collapse into itself. However, colossal amounts of energy, surging from within, unleashed by thermo-nuclear fusion, stall the collapse. When the nuclear fuel is exhausted, the star caves in. The stupendous collapse finally halts because of what is called the Pauli Exclusion Principle: Two identical electrons cannot share the same `state'. What this means is, that electrons can be squeezed in only so close to each other and no more. This is the so called `Relativistic Electron Degeneracy' pressure. The star is now earth sized, but 300,000 times more massive. So compressed, that a handful of its matter would weigh several tons! This white dwarf would slowly cool off into a `black dwarf'-sort of a cosmic coffin, the ultimate fate of all stars. Or, so scientists thought.

But Chandrasekhar had a different story. He computed that if the star has a mass, 1.44 times that of the sun, or more, the collapse would be so massive, it would over shoot the `Electron Degeneracy Pressure'. This cut off mass is the well known `Chandrasekhar Limit'. He propounded:

> "The life history of a star of small mass must be essentially different from the life history of a star of large mass. For a star of small mass, the natural white dwarf stage is an initial step towards complete extinction. A star of large mass cannot pass into the white dwarf stage and one is left speculating on other possibilities."

This could only mean that stars with a mass greater than the Chandrasekhar Limit might end up so super compressed that.... Well, the matter would completely convolute space-time around itself so that nothing, not even light, would escape from it. That is, it would end up as a `black hole', a term coined in the 60s by John A. Wheeler.

Eddington's response was not just surprising, it was shocking. Tearing Chandrasekhar's work to shreds, he averred that there was no such thing that Chandrasekhar had spoken about:

> "... the star apparently has to go on radiating and radiating and contracting and contracting until, I suppose, it gets down to a few kilometers radius when gravity becomes strong enough to hold the radiation and the star can at last find peace. I felt driven to the conclusion that this was almost a reductio ad absurdum of the relativistic degeneracy formula. Various accidents may intervene to

> save the star, but I want more protection than that. I think that there should be a law of nature to prevent the star from behaving in this absurd way."

One of these escape laws conceived by many astronomers was: Before the star could collapse into a `black hole', the shock waves of the catastrophic implosion would rip off much of the matter of the star. That is, there would be a `super nova' outburst. The remaining mass would fall below the Chandrasekhar Limit and so would peter out as a conventional white dwarf.

Eddington's outburst halted the acceptance of Chandrasekhar's new ideas for two decades. More immediately, it threatened his prospects in England. As he later recalled:

> "I had gone to the meeting thinking that I would be proclaimed for making a startling discovery. But Sir Arthur made a fool of me. I was distraught. I did not know whether to continue my career..."

After this misadventure, having taken his doctoral degree from Cambridge after having become a fellow of Trinity College when he was barely 24, Dr. Chandrasekhar had brief stints in Gottingen, Liege, Moscow and a few other places. He even tried to return to India:

> "For a whole year I tried my best to get some job, but I just could not get any. So I was forced to make the best of the tragic situation..."

Finally, Otto Struve, the celebrated exponent of stellar evolution, handpicked him, and he joined the University of Chicago and Yerkes Observatory. Here, he compiled his ridiculed work in a book and changed his field of research - a pattern, he would follow in later years. Dr. Cahndrasekhar remained there almost till the last, becoming a Professor in 1947. Thereafter, Dr. Chandrasekhar evinced interest to return to India, but was not successful. In 1953 he took up American citizenship, not through any hard feelings, as he later recounted, but through his sense of civic responsibility.

Dr. Chandrasekhar didn't subscribe to the tendency to spot and chase spectacular problems. Rather,

> "I think my motivations are different from many scientists... my motive has not been to solve a single problem but instead to acquire a perspective of an entire area..."

So, he picked up unglamorous but important topics and saw them through the end with tremendous perseverance.

He had a twelve-hour-day, seven-day-week for years on end. As the renowned astrophysicist, Martin Schwarzchild put it,

> "He just batters his way through problems no one else could do. Chandra's concentration is unbelievable-a mixture of sheer mathematical intelligence and phenomenal persistence."

Usually Dr. Cahndrasekhar sew up his researches with a bulky treatise, before switching over to a new field.

Thus, after his initial work on collapsing stars, he studied the distribution and motion of stars in galaxies and their slowing down due to the gravitational drag of neighbouring stars. Then he turned to a topic close to the heart of his illustrious uncle: the scattering of light. Though Lord Rayleigh had initiated the study in the earlier century, none had yet worked out the exact mathematics of the problem. Dr. Chandrasekhar could figure this out by the mid forties. Then magneto-hydrodynamics, and a remote topic, the stability of rotating ellipsoids and General Relativity followed in succession. In spite of open heart surgery in 1977, his "The Mathematical Theory of Black Holes" appeared in 1983. His last book, "Newton's Principia for the Common Reader" was published before his death on 21st August 1995.

Dr. Chandrasekhar was usually ahead of his times. For example, his treatise on rotating objects was appreciated, after two decades: in the light of rotating galaxies! According to Martin Schwarzchild,

> "There is not one field that he's worked in where we are not now daily using some of his results."

In later years he had once again bull-dozed into the field of black holes and cosmology, the launch point of his career. But now, bolstered by a new and bold generation of researchers, the topic was very much in vogue. In fact, when he was undergoing open-heart surgery, to his utter delight, the doctors and nurses would speak of nothing but black holes. Dr. Chandrasekhar himself considered his, "A Mathematical Theory of Black-holes," his "best," and, "most important work."

For six decades, Dr. Chandrasekhar was a giant in astrophysics, striding over different facets of the field. He had been honoured with all the top scientific awards. For example: The Bruce medal of the Astronomical Society of the Pacific in 1952. The gold medal of the Royal Astronomical Society (England) in 1953. The Rumford medal of the American Academy of Arts and Sciences in 1957. The Royal medal of the Royal Society (London) in 1962. The National medal of Science (U.S.A.) in 1967. The Draper medal of the U.S. National Academy of Sciences in 1971. And finally, the Nobel Prize in 1983.

This impressive string of awards was a tribute as much to his sustained brilliance as his dedication. For example, in the 1940s, he would drive over a hundred kilometers in the cold Chicago winters, to teach a two student class! The two pupils, T.D. Lee and C.N. Yang subsequently notched the Nobel Prize. Dr. Chandrasekhar would quip, "My whole class has won the Nobel Prize." Similarly he did yeoman service for decades, in editing the prestigious Astrophysical Journal.

Dr. Chandrasekhar and his wife-his former classmate at Madras-Lalitha, stayed in their high rise simple apartment in the Hyde Park neighbourhood of Chicago. He was active till the end. He would lament,

> "One of the unfortunate facts about the pursuit of science the way I have done it is that it has distorted my personality. I had to sacrifice other interests in life-literature, music, travelling. I've devoted all my time, every living hour practically, to my work..."

Though Dr. Chandrasekhar had been cut off from India, the country of his origin was close to his heart. He had keenly followed the development and course of scientific research in this country. In 1982, he said,

> "... the scientific atmosphere in India has enormously improved between 1961 and 1982. There is no doubt about it."

The fact is, that Dr. Chandrasekhar's loyalty to India was often misinterpreted. As he himself disclosed,

> "My loyalty is not the parochial or flat-waving type of thing. My first loyalty is to science."

In: The Best of Indian Physics
Editor: B. G. Sidharth, pp. 25-30 ISBN 1-59454-115-9


*Chapter 4*

# THE ATOMIC ENERGY PROGRAMME IN INDIA

***B. G. Sidharth***
International Institute for Applicable
Mathematics and Information Sciences, India

The idea of developing an Atomic Energy programme with an overwhelming emphasis on its peaceful utilization entered India from very early on. In fact it was Meghnad Saha who was impressed by this new emerging technology during his visit to Europe and USA in 1936. He wanted to establish a cyclotron at the Calcutta University. A student, Dr. B.D. Nag Chaudhuri, later a member of India's Planning Commission, was sent to Berkeley to learn about the cyclotron from the horse's mouth, that is from Lawrence, in 1938 itself. Saha even got Nehru, then a leading freedom fighter and later India's first Prime Minister, interested in his ideas.

Nuclear Fission was subsequently successfully achieved in 1939 and this again got Saha very excited. A few years prior to this an Indian Physicist Dr. Homi Bhabha was awarded the Rouse Ball travelling grant. On this basis, Bhabha worked with Wolfgang Pauli and Enrico Fermi in Rome. Subsequently Bhabha got the Isaac Newton studentship and this enabled him to continue studies at Cambridge and in Copenhagen, with Neils Bohr.

Homi Jehangir Bhabha came from a highly educated and cultured family with strong links to the House of Tatas and the Gandhi-Nehru nationalistic tradition. After schooling at the Cathedral High School in Bombay, Bhabha joined the Ciaus College, Cambridge, in England in 1927. Here, though he studied Engineering, his real interest was in Physics. So in 1930 he switched over to research in Theoretical Physics. Till 1939 Bhabha distinguished himself with his research in Cosmic Rays, including the interaction of these rays with matter leading to the production of electron pairs, muons and so on. When he was only 31, he was elected as Fellow of the Royal Society, in 1940. It was during this period that Bhabha interacted with the giants of Physics, like Dirac, Bohr, Pauli and Fermi. Bhabha returned to India in 1939, with the intention of going back. But due to the Second World War he had to remain behind and so joined the Indian Institute of Sciences in Bangalore as a Reader and shortly thereafter, Professor of Cosmic Ray Research. The Bhabha equation is still in vogue in the World of Physics.

While all this was going on, in 1940 the British established a Board for Scientific and Industrial Research (BSIR). Dr. S.S. Bhatnagar headed the Board and Saha was a member. The aim of BSIR was to harness Indian Industry so as to aid the Second World War efforts. In 1942 the Government formed the Council for Scientific and Industrial Research. Again Bhatnagar, Saha and others were on this Council. In 1944 Dr. Homi Bhabha approached the Tata trust for funding to set up an Institute for Fundamental Research in Nuclear Physics, a request that was complied with and thus was born the Tata Institute of Fundamental Research in 1945 with Homi Bhabha as the first Director. Curiously though TIFR began its operations in Bangalore for a few months before shifting to Bombay.

By now both Bhabha and Saha were impressing on Nehru the importance of nuclear energy for an impoverished country like India. In fact as Nehru the then head of the interim Government of India declared in 1946

> "It [is] a very grave responsibility for any country [to use] atomic bombs. A very great responsibility rests with the United States. It justified the use of [the] atomic bomb on the ground that it stopped [the] war, but by unleashing such a weapon, it [has] created a dangerous situation. The atomic bomb brought a measure of hope also. Faced with such a destructive weapon, people might wake up ... [As] long as the world [is] constituted as it [is], every country [will] have to devise and use the latest scientific devices for its protection. [I have] no doubt that India [will] develop her scientific researches and [I hope] Indian scientists [will] use the atomic force for constructive purposes. But if India [is] threatened, she [will] inevitably try to defend herself by all means at her disposal. [I hope] India, in common with other countries, [will] prevent the use of atomic bombs."

That year itself the Atomic Energy Research Committee was formed with Bhabha as the Chairman, with the aim of promoting the study of Nuclear

Physics in the academia. This committee was under the CSIR. After India became independent, under the insistence of Bhabha, a new Atomic Energy Commission was formed, de-linked from CSIR and directly responsible to the Prime Minister.

By now a shift had taken place. Nehru who had high regard for Saha and his suggestion for developing a nuclear energy programme, drifted towards Bhabha for the simple reason that Bhabha had an immediate action oriented plan rather than the more academic and thorough programme as envisaged by Saha.

In the early fifties it was envisaged that Atomic Energy in India would be, in the words of Nehru,

> ".... for social purposes. Atomic Energy represents a tremendous power. If this power can be utilized as we use hydroelectric power, it will be a tremendous boon to mankind, because it is likely to be more available and cheaper than the building of huge hydroelectric works..."

By now the Government had put in place a plan to develop India's nuclear infrastructure. However there was unease in the United States in particular. Curiously enough in July 1953, when a State owned company planned to ship sodium nitrate to China, the US Ambassador to India raised objections, citing the US Mutual Defence Assistance Act of 1951 which took a hostile view of trade in such material with the Soviet block countries including China.

By 1954 the Department of Atomic Energy had been created. However there were difficulties, particularly posed by the US in India building its own atomic reactor. After protracted negotiations and with the assistance of countries like Canada and even Britain, India's first nuclear reactor came into being. This was APSARA, commissioned in 1956.

In 1960 Indian scientists could load the Cirus Research Reactor with indigenously produced fuel rods and a few months thereafter the reactor went critical. That year it became clear that if so required India could develop an atom bomb within two or three years. But clearly, there was no intention whatsoever to do so. However the Indian Prime Minister announced that India will construct its first nuclear power station at Tarapur as also a Plutonium separation facility. When Prime Minister Nehru met the US President Eisenhower later that year, he tried to persuade the latter for a nuclear test ban. But Eisenhower could not respond positively due to the Soviet factor.

Early the next year Prime Minister Nehru while announcing India's nuclear weapons making capability within two or three years, declared at the same time that "absolutely under no circumstances shall we do so." The same year the construction of Phoenix, a spent fuel reprocessing plant began at Trombay (near

Mumbai). At that time it was projected that by 1976 India would be producing as much as three thousand mega watts of electricity.

Around this time the US State Department revealed to the US Secretary of State that China could detonate a nuclear device in 1962 itself and that, "it would be desirable if a friendly Asian power beat communist China to the punch" adding that there is "no likelier candidate than India." This suggestion was turned down.

The next year India's third research reactor Zerlina went critical. That was also the fateful year of the Chinese invasion. A section of nationalist politicians demanded a reversal of India's declared policy, so as to produce nuclear weapons. In spite of this the Indian Government stood steadfast on its principle and in fact India was one of the countries to sign the partial test ban treaty thereafter. At that time Prime Minister Nehru declared that this treaty "will take us towards disarmament and peace."

Subsequently steady progress was made particularly with the reprocessing plant. There was cooperation with the Western countries, particularly Canada and the USA.

In 1964 Prime Minister Nehru passed away and a few months later China tested its atomic bomb. The reaction in India was negative. Not just the nationalist politicians, but also some of the moderates began to demand a review of India's nuclear policy. However the new Indian Prime Minister, Sastri, refused to budge. The USA was now putting a lot of pressure on India not to respond to the Chinese test. Thus it was in 1965 that India set out to the US five conditions including guarantees for the non production of weapons. India demanded the non use of nuclear weapons against non nuclear States on the one hand and a progress towards a comprehensive test band treaty and nuclear disarmament on the other. These ideas have been a corner stone of India's nuclear policy ever since.

At that time India was attacked by Pakistan. Going against its promise to the United States not to use American weapons against India, Pakistan did so. By 1966 China was testing missiles with nuclear warheads and in 1967 it tested its first thermonuclear device. All this was the trigger for public pressure on the government to review its policy of not testing nuclear weapons. The final trigger was that in 1971 during the Bangladesh (then East Pakistan) War the nuclear powered USS Enterprise threateningly patrolled Indian waters.

The negativism in the Indian mood continued to grow over the years, particularly as the United States began reviewing its relationship with India in the light of its own cold war paranoia. The truth is that Indian sentiment was hurt. Its lofty ideals of peace, disarmament, democracy, pluralism and human values seemed lost on the world. In the Indian perception the world recognized only muscle power and money power.

Thus it was, that when India exploded its first nuclear device way back in 1974, after more than a decade of dithering, an overwhelming majority (including the scientific community) endorsed it. One of the nuclear negotiators for India, K Sundarji later General Sundarji, Head of India's Army, told me the following:

> "Before the 1974 atomic test, the American negotiators were quite overbearing. After the test there was a new respect to be found in them."

The Indian attitude to nuclear weapons has been ambivalent, reluctantly all be it. For example in 1980, the then Prime Minister, Mrs. Indira Gandhi declared that India

> "remains committed to the use of atomic energy for peaceful purposes... (But we) would not hesitate from carrying out nuclear explosions... or whatever is necessary in the national interest."

By this time the Indian Atomic Energy Program was graduating into a new era of self reliance. Indigenous heavy water plants were functioning quite well and plans for other nuclear reactors were also being firmed up.

India's peaceful nuclear program has progressed, even though the progress has been slow. The aim is to harness India's considerable Thorium resources to develop fuel efficient Fast Breeder Reactors. Plans are afoot for setting up a prototype Fast Breeder Reactor at Kalpakkam near Chennai. This would be a 500 MW project that would use the technical expertise of Indira Gandhi Centre for Atomic Research that has already set up and operated successfully a Fast Breeder Test Reactor. Meanwhile the Nuclear Power Corporation of India Limited which has generated well over nine thousand million units of electricity is already becoming profitable. There has also been some progress in the utilization of radio isotopes in medicine and other peaceful uses of nuclear energy. The fact remains that nuclear energy plays a marginal role in India's domestic power consumption scenario and is likely to remain that way for many years to come.

Unfortunately however, due to geo political reasons, India's nuclear weapons program has become potent. It is widely suspected that India has a considerable nuclear arsenal including possibly, the dreaded Hydrogen bomb. Both varieties, which were indigenously developed were successfully tested in May 1998.

India's atomic energy program, as seen above has been practically as old as that of the United States or the Soviet Union. In spite of the constraints of funds and principles, the nation has been pushed to be counted to be amongst one of the few countries with an advanced weapons program.

In: The Best of Indian Physics
Editor: B. G. Sidharth, pp. 31-38
ISBN 1-59454-115-9


*Chapter 5*

# INDIA IN SPACE

***B. G. Sidharth***
International Institute for Applicable
Mathematics and Information Sciences, India

From a fishing village to the elite club of five countries, which design, fabricate and launch their own satellites. That has been India's climb on the space ladder. As Indian Remote Sensing Satellites, now in earth orbit, scan the sub continent from nine hundred kilometers above, or Indian telecommunication satellites beam TV programmes from 36,000 kilometers above, India has achieved full-fledged space power status. But the prestige is a mere by product. What's important is, the wealth of valuable data being beamed back to selected ground stations in the country. Similarly, the newly acquired geo-stationary satellite launching capability is a boon for telecommunications and televisions.

The remote sensing data will help map out fresh areas, waste lands, potential ground water yielding sites-particularly in drought battered locations and other natural resources. In the past India was buying data beamed back from the American Landsat and from other international organizations. For some years now, roles have been reversed and agencies like NASA are buying data from India. This has enabled scientists in studies to identify potential ground water areas and to develop grasslands, forests and so on. Such data would enable advance planning and even developmental and remedial strategies. For example the success rate of water wells sunk in such pin pointed areas has been over seventy five percent. There is even the possibility of artificially recharging ground water deficient areas. For example, tanks could be built from which water would percolate into the ground or under ground dykes could be constructed.

No wonder the utilitarian space programme is a feather in the cap of Indian Science and Technology. Remote Sensing Satellites are the blue-eyed babies of the Indian space programme. The first IRS1A was launched from the Soviet Baikonur Cosmodrome on the 17th March 1988. And it enjoyed a curious distinction. It was the first ever satellite launched by the Soviets on a commercial basis. Its resolution of thirty six meters improves on the American Landsat resolution of over forty meters. This is,

> "... a remarkable achievement for the Indians, considering that this is their first attempt"

opined Dr. V.P. Opliato of the then Soviet Central Space Committee.

Aided by ground stations at Bangalore and Lucknow, IRS beamed back data to the ten-meter antenna of the National Remote Sensing Agency at Hyderabad. Regional remote sensing centres under the recently constituted National Natural Resources Management System will interpret and exploit the data. The IRS project pays itself back several times over in terms of returns.

The Indian National Satellites, INSATs have transformed telecommunications in this large subcontinent-from telephones to televisions in just about every nook and cranny.

The saga of India's bold bid to explore the universe and exploit the benefits of space applications, began in the late fifties. In 1958, the Tata Institute of Fundamental Research, Bombay, started floating constant altitude balloons, which gathered data about the atmosphere. These efforts climaxed in the formation of the Indian Council for Space Research (INCOSPAR) under the patronage of the Department of Atomic Energy.

The first major step was to launch sounding rockets. These, usually single-stage rockets, soar to a height of about 150 to 200 kilometers before plunging back. In the process, they investigate the earth's upper atmosphere and can even conduct astronomical observations. Usually the data is radioed back.

Accordingly, a rocket launching station was established at a deserted church in the fishing village of Thumba, near Trivandrum. The site was handpicked because it is almost exactly on the earth's magnetic equator - the midway curve between the earth's magnetic North and South Poles. So it is eminently suitable to explore atmospheric phenomena, particularly in the ionosphere.

The Thumba Equatorial Rocket Launching Station (TERLS) was formally launched in November 1963, when an American Nike Apache sounding rocket with an indigenous Sodium Vapour payload was shot up. Soon, boosted by U.N. sponsorship, TERLS blossomed into an International sounding rocket centre. Headed by Prof. H.G.S. Murthy, the TERLS, with American, French and Soviet help, acquired a well equipped centre, complete with launching pads, radio and radar tracking stations, payload development facilities, fabrication units and so on. Later it evolved into the prestigious Vikram Sarabhai Space Centre (VSSC), named after Dr. Vikram Sarabhai who dreamt of catapulting India on to the space map.

From Thumba, Soviet, American, British, French and India's own Rohini and Menaka rockets have been shot up. These have yielded invaluable data about the ionosphere and the upper atmosphere in general. All this will help us better comprehend the weather machinery itself. The spin off would be a more reliable forecasting mechanism: Already, monsoons are being predicted more accurately. A few weeks in advance, in fact. Apart from atmospheric and meteorological researches, solar and galactic radiations have also been investigated.

Meanwhile, the INCOSPAR was rehashed in 1969 to better manage India's growing space programme, aspirations and expertise. Consequently, the Indian Space Research Organization (ISRO) was born. In general, a space programme has four, somewhat overlapping, facets:

On the defence front, IRBMs (Intermediate Range Ballistic Missiles) and the more potent ICBMs (Inter Continental Ballistic Missiles) could be developed. These can catapult nuclear warheads across a few thousand kilometers. Satellites and space platforms armed with deadly weapons like laser and particle beams capable of razing out any given target, could be floated. As also sky-spy satellites for espionage, or for guiding naval carriers. Tasks like that.

Space science and technology constitute two other aspects. Space science focuses on the scientific exploration of the universe by blasting off space-crafts to neighboring worlds, by launching satellites outfitted with observational instruments and so on. Space technology concentrates on industry and technology oriented experiments in outer space. For example, processing of materials in conditions of weightlessness. All this could well transform future technology.

Finally, `space applications' utilizes satellites in diverse areas with imminent benefits. Survey satellites outfitted with remote sensing devices can scan and monitor the various resources of the earth like vegetation, water and mineral wealth: In fact, infra-red and microwave sensors are priceless tools in Agriculture, Hydrology, Geology and related fields. Moreover, satellites aid meteorological studies while communication satellites are crucial in telecommunication, including TV, Telephone and Radio transmission over vast areas.

Obviously, for a developing country like India, the accent is very much on space applications. Other aspects receive a much lower priority.

ISRO monitors India's space programme, from several centres. For example, the Vikram Sarabhai Space Centre (VSSC) near Trivandrum, encompasses, apart from TERLS, a space research and development complex, equipped with facilities to produce and develop scientific payloads and instruments, propellants and rockets - both sounding rockets like the Rohini and Menaka series and also the more powerful satellite launchers.

The ISRO satellite centre at Peenya, Bangalore, develops and fabricates all Indian satellites.

The Sriharikota Rocket Range (SHAR) located on the East coast of Andhra Pradesh, a hundred kilometers from Chennai, can boost both multi-stage sounding rockets as also full blown satellite launch vehicles.

The Space Applications Centre (SAC), Ahmedabad, Western India, as the name suggests, has developed telecommunication and TV transmitting and receiving sets, remote sensing equipment for earth survey satellites and so on.

The 19th of April, 1975, was a breakthrough day for India's space programme. The first satellite, developed at the satellite systems project, Bangalore and christened after the renowned fifth century Indian astronomer-cum-mathematician, Aryabhata, was blasted away from a Soviet cosmodrome, by an Intercosmos rocket. The 1.59-meter by 1.19 meter, 358.5 kilograms, quasi-spherical satellite, was covered with 18,500 blue and violet solar cells. These transformed sunlight into electricity. Further, a nickel-cadmium battery stored energy for use when the satellite would duck into the dark earth-shadow zone.

Aryabhata soared to an altitude of 594 kilometers along a nearly circular, 96 minute orbit, looping above North Western India. The satellite beamed back data on its own functioning and received commands from the Sriharikota ground station, operating for more than the initially budgeted six months. The project exhibited India's capability in fabricating satellites as also in tracking and reception techniques and transmission of commands.

On the seventh of June, 1979, another Soviet Intercosmos rocket punched India's second Satellite, Bhaskara, named after the prodigious eleventh century Indian mathematician-cum-astronomer. The 444 Kilogram Bhaskara, deceptively similar in structure to Aryabhata, zoomed into a similar, 95 minute, orbit between 557 and 512 kilometers. But its function was totally different. Outfitted with TV cameras and an ISRO-developed, Satellite Microwave Radiometer (SAMIR), it was exclusively an Earth Observation Satellite. The microwave radiometer operated smoothly, transmitting priceless data. Initially the TV cameras did not function. As late as May 1980, one of the cameras started responding and relaying views of vast tracts of the country.

Meanwhile, India's programme to harness satellites for telecommunication forged on. It originated in 1965 with the first Experimental Satellite Communication Earth Station at Ahmedabad. The success of the Satellite Instructional Television Experiment (SITE) in 1975-1976, further boosted these efforts. SITE employed the American Comsat ATS-6 that beamed educational programmes. A newly commissioned earth station at Delhi and low cost Direct Reception Systems installed at 2,400 widely strewn villages received and broadcast these programmes. The next step was STEP - the Satellite Telecommunication Experiment Project: This experiment hooked up the Franco-German satellite, Symphonia, with ground stations at Ahmedabad, Delhi and Madras, during 1977-79. A notable innovation was the injection of the Mobile Transportable Communication Terminal that established links with other earth stations to telecast live, events in remote places. STEP was conceived to provide experience for India's envisaged domestic satellite communication network, employing the US built INSAT (Indian National Satellite).

And then, in 1980, the European Space Agency's experimental three-stage Ariane rocket blasted off the telecommunications satellite APPLE, from French Guinea in South America. APPLE (Ariane Passenger Payload Experiment) had been developed at ISRO's Bangalore wing. It was successfully maneuvered into a geostationary "parking slot" above Indonesia: India became the fifth country to achieve this. Apple functioned smoothly.

In 1982, INSAT 1A, built for India by the Ford Aerospace Communications Corporation, was catapulted by the American two-stage Delta rocket. Again, it was successfully steered into its geostationary parking slot. But the 50,000 solar-cell-solar-sail, the key power supplier during the satellite's projected 7-8 year life span, did not flap open. Reluctantly, INSAT 1A was abandoned. As a stopgap, India leased out transponders from the U.S. INTELSAT organization and the Soviet Statsionar 6, for telecommunications and the national TV network.

The focus was now on the standby INSAT 1B. The U.S. Space Shuttle, Challenger, ejected it into earth orbit in August 1983, almost exactly a year after INSAT 1A's launch. INSAT 1B's solar sail also threatened to go the INSAT 1A way. But frantic efforts by the Master Ground Control in Hasan, Karnataka cleared the hitch. INSAT 1B was hooked on for long distance telephone facilities, for beaming TV programmes, aided by 13 high power and 112 low power transmitters, for radio broadcasts by All India Radio and for meteorological studies. Propped up by the crucial Meteorological Data Utilization Centre in New Delhi and twenty secondary centres, the Meteorological department not only provided regular weather reports but also advance disaster warnings. Indeed, India had gatecrashed into a new era in telecommunications.

Meanwhile, on the 18th July, 1980, India's four-stage, solid fuel rocket, SLV3 successfully blasted the 40 kilogram Rohini satellite (RSI) from Sriharikota and India zoomed into the elite space club, trailing the USSR, USA, France, China and Japan. The SLV3 could eject a forty-kilogram satellite into an elliptical earth orbit between 200 and 800 kilometers. RSI collected vibration data, spin rate of the fourth stage, and velocity and other parameters of the SLV3 vehicle during the twelve minute ascent and launch phase. Subsequently two other Rohinis were successfully punched up from Sriharikota.

Yet another crescendo in India's space saga was Squadron Leader Rakesh Sharma's odyssey in April 1984. Shot up by a Soyuz T spacecraft, Sharma spent a week aboard the orbiting space station, Salyut 7, performing various experiments.

A sixty-five meter launcher was developed to punch the more potent Polar Satellite Launch Vehicle (PSLV). Such a rocket could catapult a Remote Sensing Satellite into an orbit that would loop over the North and South Poles. When it was completed, India could boast of being a truly full fledged space-power-fabricating and launching its important satellites.

Also on the drawing boards were plans to build liquid propellant rockets and to develop within ten years, powerful launch vehicles which could float upto 100 kilogram multi-purpose satellites - for remote sensing, telecommunications and so on. These aspirations were realized with the development of, first, the PSLVs, the Polar Satellite Launch Vehicles for launching Remote Sensing satellites and subsequently, in the nineties, the GSLVs with cryogenic engines with the capability of putting telecommunications satellites in a 36000 kilometer high geo-stationary orbit.

These larger rockets could be re-hashed into the ominous IRBMs and ICBMs. India is equipped too with the deadly nuclear warheads. However, crippled with fund crunches, India had tinkered out a blueprint which was well conceived with optimum utilization of meager finances. India's space pioneer, the late Dr. Vikram Sarabhai then succinctly put it,

> "We do not have the fantasy of competing with the economically advanced nations in the explorations of the Moon or the planets... But we are convinced that if we are to play a meaningful role nationally and in the community of nations, we must be second to none in the application of advanced technologies to the problems of man and society, which we find in our country."

Nevertheless owing to growing pressure from the scientific community and a growing economic clout, India is now venturing out even further with the proposed moon mission, Chandrayaan 1, possibly the first of a series. (Interestingly, India's space budget far exceeds that of Russia and Britain). This mission aims to put a space craft into a lunar polar orbit at an altitude of about 100 kilometers, sometime in 2007-2008, using an Indian space craft and an Indian launch vehicle. It is expected that the space craft would have a life time of about two years. It would carry out high resolution mapping of topographic features in three dimensions and of the distribution of various minerals and elemental chemical species including radioactive nuclides covering the entire lunar surface using a set of remote sensing payloads. Other objectives include:

- High resolution mineralogical and chemical imaging of permanently shadowed north and south polar regions.
- Search for surface or sub-surface water-ice on the moon, specially at lunar poles.
- Identification of chemical end members of lunar high land rocks.
- Chemical stratigraphy of lunar crust by remote sensing of central upland of large lunar craters, South Pole Aitken Region (SPAR) etc , where interior material may be expected.
- To map the height variation of the lunar surface features along the satellite track.
- Observation of the X-ray spectrum greater than 10 keV and stereographic coverage of most of the moon's surface with 5 m resolution, to provide new insights in understanding the moon's origin and evolution.

Chandrayaan 1 would be in the nature of a technology demonstrator.

The programme would also, it is hoped, revitalize interest in Science and Technology.

It is not crowded up there. Only intensely competitive. And India, wriggling to extricate one foot from the jaws of prehistory, has planted the other firmly into outer space: She has stuck her neck out to harness its spin offs.

## References

[1] B.G. Sidharth, *"The Celestial Key to the Vedas"*, Inner Traditions, Vermont, USA (1999).

[2] N. Sinha, *"The Vaisesika Sutras of Kanada"*, S.N. Publications, New Delhi (1986).

[3] G. Venkataraman, *"Raman And His Effect"*, Universities Press, Hyderabad (1995).

[4] G. Venkataraman, *"Saha And His Formula"*, Universities Press, Hyderabad (1995).

[5] C.N. Srinivasiengar, *"The History of Ancient Indian Mathematics"*, The World Press Private Ltd., Calcutta (1967).

[6] G.R. Kay, *"Hindu Astronomy"*, Cosmo Publications, New Delhi (1981).

[7] G.G. Joseph, *"The Crest of the Peacock"*, Affiliated East-West Press Pvt. Ltd., New Delhi (1995).

In: The Best of Indian Physics
Editor: B. G. Sidharth, pp. 39-56
ISBN 1-59454-115-9


*Chapter 6*

# RESURGENCE OF MODERN SCIENCE IN INDIA - A NATIONAL MOVEMENT

***Kankan Bhattacharyya***
Department of Physical Chemistry,
Indian Association for the Cultivation of Science

India has a rich tradition of science and technology. The Indus valley civilization (Harappa and Mahenjodoro) was among the most advanced civilizations of its time. The first use of bricks, town planning, hydraulic port, smelting furnaces and many other features of Harappa and Mahenjodoro demonstrate that reasonably advanced scientific and technical knowledge was available in the Indian sub-continent as early as in 2500 BC. The Indian supremacy in science and technology continued for many centuries. In 1963, the noted biologist and historian of science J. C. Needham, FRS reminded the students in his annual lecture as a Master of Gonville and Caius College, Cambridge,

> "It is good to remember that our pious founders were not the only men, and that Christendom is not the only culture to set on foot great institutions of learning. When the men of Alexander the great came to Taxila in India in the fourth century BC, they found a university there the like of which had then been seen in Greece and was still existing when the Chinese pilgrim Fa-Hsien went there about 400 AD."

There are many reasons why India could not retain its scientific tradition. Needham wrote,

> "Theoretical materialism in ancient India had been obscured and vilified by all Indian religions. As a result, medical men (Charak) had to fight against religious theories."

According to a religious theory a disease occurs because of sins committed by a person and may be cured not by medicines but by worshiping god. Such superstitions led to the decline of the medical research initiated by Charak and Shusrut.

The caste system was another major impediment. P. C. Ray commented,

> "In ancient India, useful sciences perished with the institution of the caste system."

Following casteism, only higher caste people were allowed to read. The higher caste people never did any thing using hands. Thus their bookish knowledge was of no material benefit to the society. The artisans, farmers, carpenters blacksmiths who actually used many instruments had no access to knowledge. In Ramayana, Rama beheaded Shambuk (a shudra) for reading the shastras. The so-called "educated" higher caste people never appreciated the technical skills of our artisans. In Indian folk stories, the lower caste people and the artisans are ridiculed for their ignorance (e.g. the weavers were the proverbial fools). The women were debarred from education. Since the educated intellectuals never actively participated in any manufacturing and fabrication process, the production methods did not improve in India. About this M. N. Saha wrote

> "Casteism destroyed the relation between head and hands in India. We belittle anything done using hands. For this reason, there has been no technological innovation in India for many centuries."

Perhaps, the people belonging to the higher caste led by the priests feared if education spreads to all sections of the society, their supremacy would be threatened. Because of their conceited ideas India remained scientifically and militarily backward and was defeated time and again by foreign invaders.

The British captured power in India in 1757. Around 1800, when the industrial revolution was sweeping across the whole of Europe, the demand for western education began to crystallize among the elite of the Indian society and also among a few Englishmen. In 1811, Lord Minto observed,

> "It is a common remark that science and literature is in a progressive state of decay among the natives of India. The number of the learned is not only diminished, the circle of learning appears to have contracted. The abstract sciences are abandoned, polite literature neglected and no branch of learning cultivated but that is connected with the peculiar religious doctrines."

Before 1800, the East India company had started only two educational institution in India, the Calcutta Madrasa (1781) and the Benaras Sanskrit College (1791). In both the places, students were taught traditional Muslim and Hindu philosophy. Western education was not available in both the places.

Raja Ramamohan Ray (1772-1833) a religious reformer and founder of the Brahmo religion may be considered as the harbinger of modern thought and scientific education in India. In 1815, he and David Hare mooted the idea to start a college with facilities for western education in Calcutta. David Hare was an uneducated watch-maker and a bachelor who spent most of his life in Calcutta with a missionary zeal to spread western education in India. Rammohan's idea to found a modern college was accepted by many. However, a section of the Hindu leaders said that they would donate and participate only if Rammohan leaves the committee. Rammohan stepped aside and the Hindu leaders founded the Hindu College in 1817.

In 1824, the British government founded the Sanskrit College in Calcutta. Raja Ramamohan opposed this. In a long letter written to Lord Amherst, Ramamohan asserted

> "We find that the government is establishing a Sanskrit school under Hindu Pundits. This seminary (similar in character to those which existed in Europe before Lord Bacon) can only be expected to load the minds of youth with grammatical niceties and metaphysical distinctions, of little or no practical use. If it had been intended to keep the British nation in ignorance of real knowledge, the Baconian philosophy would not have been allowed to displace the system of schoolmen, which was the best calculated method to perpetuate ignorance. In the same manner Sanskrit system of education would be the best calculated to keep this country in darkness. But as the improvement of the native population is the object of the (British) government, it will consequently promote a more liberal and enlightened system of instruction, embracing mathematics, natural philosophy, chemistry, anatomy with other useful sciences."

The British government did not accept Rammohan's idea to introduce western education, immediately. In 1828, a 19 year old young man Derozio (1809-1831) joined Hindu College as a lecturer. Derozio's teaching instilled a tremendous urge among the students to acquire western knowledge. About this period P. C. Ray wrote,

> "There was a ferment all around. A new world had been opened out: new aspirations were awakened. Roused from a period of stu-

por and stagnation Young Bengal began to realize that there are immense possibilities in the Hindu nation."

The liberal and progressive ideas introduced by Rammohan, Derozio and others evoked a strong reaction from the conservative section of the Indian society. A severe controversy arose between the Anglicists (Ramamhon, Derozio, Hare and others) and the Orientalists. The Orientalists felt western education, particularly teaching in English language will destroy the Indian culture. Finally, the Anglicists won because of the support of Lord Macaulay, who came to India in 1834.

Macaulay echoed Rammohan's views in his famous minute on education in India. He recommended use of English as a medium of instruction and promotion of western education. Macaulay's minute was officially accepted by Lord Bentinck in 1835. It still took 20 more years for the British government to take over the Hindu College and convert it into Presidency College in 1855. Noting the persistent efforts of Rammohan and others, Duff commented that "English education was forced on the British government."

In 1835, the Medical College was established in Kolkata. On Jan 10, 1836, Madhusudhan Gupta became the first Indian to dissect a human body at the Calcutta Medical College. Thus, the medical tradition of Charak and Shusrut was revived.

In 1851, Iswarchandra Vidyasagar (1820-91) joined as a Principal of Sanskrit College. He radically reformed the Sanskrit College. He was of the opinion that the basic aim of education is to teach the students to think logically and express clearly. In his notes to the education department he commented that the Indian philosophy had become outdated and did not teach the students to think logically. He felt that the students should read both Indian and western philosophy to understand our strengths and weaknesses. Therefore, he introduced Mill's book on western (inductive) logic in Sanskrit College. He also prescribed that Mathematics should be taught in English language instead of the prevailing practice of using complicated rhymes. He asserted that the mathematical statement must be precise and straight forward. According to him knowledge of Sanskrit is necessary to beautify our language but knowledge of English is essential to make our expression precise and crisp. He started more than 50 model schools under Sanskrit College. But these reforms were not liked by the British Government and in protest Vidyasagar resigned in 1858.

The intellectual movement initiated by Rammohan, Derozio and Vidyasagar eventually bore fruit. Radhanath Sikdar, a direct student of Derozio and an employee at the Trigonometric Survey of India carried out a detailed research work on the heights of different peaks of the Himalayas using the available data.

In 1852, he arrived at the conclusion that the so called peak XV is the highest mountain of the world. This peak was named Mount Everest after the then chief of the Trigonometric Survey of India. S. N. Bose (discoverer of the Bose Statistics) hailed Radhanath's work as the first original scientific and technical discovery by an Indian in the modern era.

In 1857 the Sipahi Mutiny, the first rebellion against the British rulers took place in India. By this time, the British government realized that something must be done to appease the Indian society. Thus in 1855, they took over the Hindu College in Kolkata and renamed it Presidency College and made arrangements for science education. The Madras Presidency College started at the same time. In 1857 the government established the Calcutta, Madras and Bombay Universities.

Around 1870, Dr. Mahendra Lal Sircar (1833-1904) a medical practitioner started a campaign for an original scientific research institute in India. Sircar, founded the Indian Association for the Cultivation of Science (IACS) in 1876. In 1876, Raman and Ramanujan were not even born and the other pioneers of Indian Science, J. C. Bose, P. C. Ray and Asutosh Mookerjee were in their teens. At that time, there was no serious desire among the Indians to do scientific research and the colonial British rulers were not interested to sponsor scientific research in India.

It is obvious that M. L. Sircar was a visionary and was very much ahead of his times. His views and ideals are eloquently expressed in his many addresses as the Secretary of IACS at its annual meetings. He was imbibed with a deep sense of nationalism. He tried to get funds from the British government but failed. IACS did not receive any government grant in the first 50 years of its existence. Instead of complaining about it, he maintained a very cordial relation with the British government. But he very cautiously kept IACS free from all government influences.

Mahendra Lal said

> "The sole function of the Association (IACS) will be science-learning and science-teaching. We should carry on unaided by (British) government or more properly, speaking, with out seeking its aid. I want freedom for the institution. I want it to be solely native and purely national."

Mahendra Lal and Father Lafont (of St. Xaviers College) organized many scientific lectures at IACS. Among our pioneers, J. C. Bose and Asutosh Mookerjee were associated with IACS as teachers and P. C. Ray was a student.

Sircar built a laboratory and library with the donations received from many philanthropic Indian aristocrats (Maharaja of Vijianagram, Kalikrishna Tagore and others). Unfortunately, original research did not start at IACS in Sircar's lifetime. Evidently Sircar was very unhappy about it. In 1902, in the annual meeting of IACS he lamented

> "I do not know how to account for this apathy of our people towards the cultivation of science. And therefore I am forced to confess that I made a mistake in starting the project of founding a Science Association at all and that I have wasted a life as I have told you, in attempting to make it a national institution."

In the annual meeting of 1903, he sent from his death bed his last address as a secretary in form of a written speech in which he said,

> "I reiterate my conviction that if our country is to advance at all, it can only be by means of science. To this end, I have given the best part of my life, but I am sorry to leave this world with the impression that my labours have not met with the success it deserves."

During this period a young marathi girl, Ana Pandurang provided some consolation and support to Sircar. Ana addressed Sircar as "Baba" (father). In 1879, she wrote in a letter,

> "Dear Baba, the world is cruel and slow to recognize worth. But do not suppose that your life and labour has been wasted. You may never be satisfied but the fruit will remain and though you may doubt it now, men will yet rise up to call you blessed."

Sircar's dream and Ana's prediction were fulfilled by C. V. Raman who initiated original scientific research at IACS in 1907. He discovered the Raman effect at IACS in 1928 for which he was awarded the Nobel Prize in 1930. At a civic reception after the Nobel Prize, Raman said,

> "To two men (M. L. Sircar and Asutosh Mookerjee), especially, I owe a debt of gratitude that can never be repaid. It was the late Dr. Mahendra Lal Sircar, who, by founding the Indian Association for the Cultivation of Science, made it possible for the scientific aspirations of my early years to continue burning brightly. Dr. Sircar devoted a lifetime of labour to the institution he created and equipped in the hope that it would some day be utilized for the advancement of science in India. Its doors were open, awaiting the arrival of someone. That arrival happened to be myself. Dr. Sircar did not, alas live to see his aims accomplished. It is IACS which first afforded me opportunities for scientific research."

The modern practice of scientific research and publishing in journals was started in India by Sir Asutosh Mukherjee (1864-1924). Asutosh was an extremely gifted Mathematician. He published his first paper in 1881 when he was in his teens. He published nearly 20 papers and became a fellow of the Royal Society of Edinburgh. He wanted to become a Professor of mathematics in Calcutta University. The then vice-chancellor Gurudas Banerjee failed to create a post for him. So Asutosh shifted to law which was a very fashionable profession in those days. He had two doctorate degrees- one in law and the other in mathematics. He became a Professor of Law at the Calcutta University and later, Chief Justice of the Calcutta High Court. Asutosh remarked

> "I began my life as a research student in mathematics when research was practically unknown in this country. The mission of my life was to be a research professor in my university. Gurudas Banerjee, the vice-chancellor, made desperate attempt to create a chair for me but failed. I drifted to law but I made a determination that I would devote myself so that in the next generation any aspiring scholar may have adequate opportunities."

We will describe later how Asutosh fulfilled his dreams.

In the 1880-s three Indians returned to India after studying in England. The first among them was a Geologist, P. N. Bose (1855-1934). He secured third rank in his graduation examination at the London University and joined the Royal School of Mines. He was involved in politics and often criticized the government. The government packed him off to India with a job in the Geological Survey of India (GSI). He eventually became the senior-most officer of GSI and was all set to become its first Indian Director. But the British Government never wanted an Indian to head GSI and hence, Bose was superseded by one of his junior to the post of Director. In protest, Bose resigned from GSI and joined the State of Mayurbanj as a State Geologist. When J. N. Tata (1839-1904) was planning to set up a steel plant in India it was P. N. Bose who advised him to set it up at Jamshedpur.

Sir Jagadish Chandra Bose (1858-1937) returned to India in 1885 and was appointed as a Professor of Physics at the Presidency College Calcutta. In those days there were two classes of Professor at the Presidency College. The Europeans enjoyed a higher salary and greater status. J. C. Bose resented this in an original manner. He boycotted the salary. So in a way J. C. Bose was the pioneer in satyagraha (peaceful non-cooperation movement) which Mahatma Gandhi adopted much later. The British government relented finally in 1888 and Bose was the first Indian to enjoy a similar status (Imperial Service) to a European Professor.

J. C. Bose's father ventured into many indigenous industries and incurred heavy financial losses. However, Bose never considered his father a failure. He used to visit the Swadeshi Melas and proudly lectured on his father's "failures." He said about his father,

> "A failure? Yes, but not ignoble nor altogether futile. His failure contributed to the success of many."

J. C. Bose was a great innovator. At Presidency College, Bose started working on the reflection, refraction and transmission of millimetre (mm) waves. He ingeniously devised diffraction gratings for the mm waves by inserting tin plates in railway timetables at regular intervals (after equal number of pages). In order to rotate polarization of mm waves he used twisted jute bundles. In 1895, he demonstrated wireless communication at Calcutta. Using mm waves, he rang a bell and exploded some gunpowder which was placed in a room at a distance and separated by more than one wall. In 1896 the Daily Chronicle of England reported: "The inventor (J.C. Bose) has transmitted signals to a distance of nearly a mile and herein lies the first and obvious and exceedingly valuable application of this new theoretical marvel." The first successful wireless signaling experiment by Marconi on Salisbury Plain in England was demonstrated in May 1897. The 1895 public demonstration by Bose in Calcutta predates this experiment. On an invitation by Lord Rayleigh, in a Friday Evening Discourses on Jan 29, 1897 Bose reported on his microwave (millimeter-wave) experiments at the Royal Institution. Unfortunately, Marconi was considered to be ahead of J. C. Bose and was awarded the 1902 Nobel Prize in Physics. However, recent research has unequivocally established that J. C. Bose did it much before Marconi. In an article Emerson concluded that,

> "It appears that Bose's demonstration of remote wireless signaling has a priority over Marconi. He (Bose) was the first to use a

semiconductor junction to detect radio waves, and he invented various now commonplace microwave components. Outside of India, he (Bose) is rarely given the deserved recognition. Further work at millimeter wavelengths was almost nonexistent for nearly 50 years. J.C. Bose was at least this much ahead of his time."

Around 1900 Bose shifted to biology and made pioneering contributions on responses of plants to electric impulses and on photosynthesis. He summarized his work in these areas in several books e.g. "Physiology of Photosynthesis" (1924), "The Nervous System of Plants" (1926) etc. He founded the Bose Institute in 1917 and became a fellow of the Royal Society (FRS) in 1920.

P. C. Ray (1861-1944) joined Presidency College as an Assistant Professor in 1889. He initiated original research in chemistry and established a school of students, founded the Indian Chemical Society and a Journal. He was also a pioneer in indigenous chemical industry. In 1901, he founded the Bengal Chemical Company. Ray had a keen interest in history of science, particularly chemistry. He was a nationalist and an ardent follower of Gandhiji's Khadi (indigenous technology) movement.

In 1905, as a part of their "divide and rule policy" the British government partitioned Bengal on the basis of religions- Hindu and Muslim. People belonging to both the religions fought unitedly against this decision. The anti-partition movement had a great impact on Indian youth. Young Meghnad Saha (1893-1956) joined this movement and was expelled from his school in 1905. Finally, the British had to reverse the decision and Bengal was re-united in 1912. About this period S. N. Bose wrote "The national movement in 1905, aroused a strong desire for economic self reliance among the Indians. Science education was inadequate. The leaders decided to establish our own university so that the students would build a strong and self-reliant nation. Thus, in order to uphold our prestige a war was declared against the foreign rulers." The National Council of Education came up as a result of this. Two Indian lawyers Sir T. N. Palit and Sir R. B. Ghosh donated nearly 25 lakhs of rupees in 1914 for this.

Two great Indians died in 1904. One of them was Dr. M. L. Sircar and the other was J. N. Tata. Tata died when four of his dream projects were taking shape. First, the Taj Hotel had just came up at Bombay. Second, the Tata Steel company was being founded at Jamshedpur. The third one was a hydro-electric project. Tata's fourth project was to found a big research institute for which he left aside about half of his personal property worth Rs. 30 lakhs. With this money the Indian Institute of Science came up in Bangalore in 1911. Bangalore was chosen for its climate and also the benevolence of the Maharaja of Mysore. The Maharaja donated 300 acres of land, rupees 5 lakhs for the building and one lakh as annual grant.

During the turbulent period of the anti-partition movement in Bengal, Asutosh Mukherjee became the vice-chancellor of the Calcutta University in 1906. In the same year C. V. Raman (1888-1970) stood first in the M. Sc. Examination of the Madras University without attending the classes. About this Raman said,

> "I enjoyed a measure of academic freedom which seems incredible. To mention only one in detail, during the whole of my two years of work for the MA degree, I remember attending only one lecture."

Through out his life Raman gave more importance to research over conventional class room lectures. Thus when he was appointed as a Professor of Calcutta University, he forced the University to take a decision that "Mr. Raman will not be required to undertake any teaching work in the MA and M.Sc. classes to the detriment of his research or his assisting advanced students in their research work."

While he was an M.Sc. student, Raman published three papers unaided by anybody. Raman sat for the Indian Financial Service and ranked first. In 1907 Raman came to Calcutta as an assistant auditor-general of the Government of India. On reaching Calcutta (Kolkata) Raman rented a house at Scots Lane which is very close to IACS at 210 Bowbazar Street. Very soon Raman discovered the institute (Indian Association for the Cultivation of Science). A. L. Sircar (M. L. Sircar's son) welcomed Raman to carry out research in his spare time. Within a short time Raman started publishing from IACS. In the annual meeting of IACS in 1907 A. L. Sircar proclaimed

> "The prophecy of the great man (M. L. Sircar) is going to be fulfilled. If circumstance do not go against us, Raman will be the brightest ornament of this Association."

There could not be any statement more prophetic. Raman continues to be the brightest ornament of not only IACS but the entire Indian science.

During this period Asutosh adopted a very clever role. He did not go into confrontation with the British Government. He slowly tried to build the Calcutta University as an outstanding centre of learning. In 1908, he established the Mathematical Society. Then he persuaded the nationalist leaders to hand over the entire fund of the National Council of Education. With this money and the two houses donated by T. N. Palit, Asutosh founded the University College of Science, the Science Department of Calcutta University in 1914. In the same year the Indian Science Congress was founded and Asutosh was elected its first President. Asutosh tried hard to get a matching grant from the British

Government but he failed. The British Government even boycotted the foundation ceremony of Science College. 4 days after the foundation of Science College, Asutosh's term as vice-chancellor expired and his term was not renewed. Though Asutosh did not remain the vice-chancellor, as the President of the Palit Trust he controlled the appointment of teachers in the Science College. One of the conditions of the Palit Trust was that all the teachers of the Science College will have to be Indian in origin. It was very difficult to find good Indian teachers.

Before discussing how Asutosh found the teachers and fulfilled his dreams we will digress a little. In 1908, Srinivas Ramanujan (1887-1920), the mathematical genius failed in all subjects in FA (pre college) except mathematics. However, within three years, in 1911, Ramanujan published his famous paper on Bernoulli's number. It immediately caught the attention of Professor Hardy and on his invitation Ramanujan sailed for England in 1913. Ramanujan got his B. Sc. (equivalent to Ph.D.) from the Cambridge University in 1916. In 1918 he was elected a FRS. Sadly, Ramanujan passed away in 1920 leaving behind a diary filled with his ideas and numerous unfinished works.

About this period the famous Indian astrophyisicist and later Nobel Laureate S. Chandrasekhar (1910-95), a nephew of C. V. Raman commented

> "Between 1920 and 1925 we suddenly had 5 or 6 internationally well-known men. It was a part of the national movement to assert oneself. India was a subject country, but in the sciences we could show the West in their own realm that we are equal to them."

We now come back to Asutosh's difficulties in starting the Science College. As the Palit Professor of Chemistry and Physics Asutosh chose P. C. Ray and C. V. Raman, respectively. Ray agreed to become a Professor without any salary. Raman left his lucrative government job and accepted the Professor post which carried a salary nearly one half of his Government job. Moved by the sacrifice of Raman, Asutosh said at the Senate meeting of the Calcutta University

> "I feel for the courage and spirit of self-sacrifice with which Raman has decided to exchange a lucrative official appointment for a University Professorship, which does not carry even liberal emoluments. This one instance of self-sacrifice encourages me to entertain the hope that there will be no lack of seekers after truth in the Temple of Knowledge which it is our ambition to erect."

Raman, however, created some several difficulties. First, there was a condition that the Palit Professor had to have a foreign training. Raman did not have a foreign training and refused to go abroad for foreign training. So the Senate of Calcutta University recorded

> "Mr. Raman's acceptance of the offer of the Palit Professorship on condition that he will not be required to go out of India."

About the foreign training, during a civic reception given by the Mayor of Calcutta (Dr. B. C. Roy) after Raman's Nobel Prize, Raman said

> "I never had training in foreign laboratories. I believe this was fortunate, for the highest scientific inspiration is that which comes from within oneself. I do not suggest that we have nothing to learn from Europe or America. But surely it is better that we learn to accomplish within our boundaries."

The second problem was what we discussed earlier that Raman refused to have an official teaching responsibility. Raman however, did teach at the Calcutta University. Ramdas, who attended Raman's M. Sc. classes at the Calcutta University, wrote,

> "Though under the terms of his appointment, he (Raman) was entirely free from any teaching responsibility he was equally enthusiastic to take a part in M.Sc. teaching. Raman once remarked that the best way for him to master or revise any subject was indeed to lecture on it. Often he used to take the entire forenoon for 2 and some times even 3 hours - such was his tremendous love for teaching."

Raman's initial reluctance to teach and also the lack of sufficient number of qualified teachers put Asutosh in serious difficulty to appoint teachers at the Science College. At this stage, a group of very brave and brilliant students approached Asutosh to give them a chance. They passed the M. Sc. Examination of the Calcutta University in 1915 by studying the science subjects at the Presidency College. They included S. N. Bose and M. N. Saha (first and second in the department Mixed Mathematics), S. N. Ghosh (first in Physics) and J. C. Ghosh (first in Chemistry). Asutosh immediately recognized their talent and selected them as lecturers. They were given 2 years' time to procure textbooks, build the laboratories and prepare the courses. This was a very difficult task at that time. The reason is that the first World War started in 1915 and as a result, imports of all books and instruments were totally stopped. So these young men started hunting the many personal collections of many rich and educated people in India for books and instruments.

However, before classes began, the Physics teaching at the Science College suffered a severe jolt. Two lecturers of Physics- S. N. Ghosh and J. Seth were debarred from teaching because of their connections with the freedom movement. Seth was put under house arrest. S. N. Ghosh escaped from India to avoid arrest. Professor D. M. Bose a Professor of Physics went to Germany in 1915. The Germans did not allow him to come back. So there was virtually no body to teach Physics. This is why two young mathematicians, S. N. Bose and M. N. Saha started teaching Physics at the Science College.

The first world war did one good thing for Indian science, by not allowing P. C. Mahalanobis to go back to England. Mahalanobis got a Tripose from the King's College, Cambridge in 1915 and got a fellowship to do a Ph.D in Physics with Professor C.T.R. Wilson (the 1927 Nobel Laureate). When the war broke out he joined the Presidency College in Calcutta as a Professor of Physics. Though he had a basic background in Physics he was one of the international pioneers of modern statistics and sample survey techniques. He founded the Indian Statistical Institute and started the famous journal "Sankhya." After Independence, he took a leading role in the Planning commission. The second 5-year plan (1957-62) and the subsequent stress in establishing heavy industries is often described as the Mahalonobis Model.

Raman joined Science College in 1917. By this time M. N. Saha had published his first two papers. Saha and Bose translated the original papers on relativity written by Einstein and Minkowski from German to English. Calcutta University published this with a forward by Mahalanobis. Saha published his famous paper on thermal ionization of stars around 1920. In this paper he combined, quantum and statistical mechanics with Chemical equilibrium and atomic spectroscopy. His work explained the results collected in many observatories for many decades. His method led to the determination of the temperature of a star with an error of a few per cent. The previous methods had an error of 100-1000%. Eddington ranked Saha's work as one of the 15 most important discoveries in Astrophysics. Saha moved to Allahabad University in 1923.

In 1917, after joining the Science College Raman observed that

> "A real school of Physics has grown up in Calcutta, the like of which does not exist in any other Indian University and which even now will not compare unfavorably with those in European and American Universities."

Raman became an FRS in 1924. M. N. Saha became an FRS in 1927.

In 1921, the Dacca University was founded. S. N. Bose and J. C. Ghosh moved from Calcutta to Dacca.

S. N. Bose discovered the Bose Statistics in 1924 when he was at Dacca. It is said that S. N. Bose's paper was initially rejected by the Philosophical Magazine. Bose sent the paper to Einstein. Einstein arranged to publish it in Zeitschrift Fur Physik with a note that this work is "an important step forward." Around the same time, Langevin sent to Einstein a Ph. D. thesis submitted by De Broglie. Einstein combined both Bose's theory and De Broglie's hypothesis to develop the quantum theory of gases. Schrodinger later said that the discovery of Wave-mechanics and the Schrodinger equation was largely influenced by Einstein's paper.

It is true that Bose did not get a Nobel Prize but his name is permanently engraved in the vocabulary of modern physics in the form of the phrases Bose Statistics, Bose Condensation and Bosons. In order to emphasize the impact of Bose's work one may quote the press release of the 2001 Nobel Prize in Physics. The press release said the Nobel Prize was awarded

> "for the achievement of Bose-Einstein condensation in dilute gases of alkali atoms, and for early fundamental studies of the properties of the condensates."

The report further stated,

> "This year's Nobel Laureates have succeeded in discovering a new state of matter, the Bose-Einstein condensate (BEC). In 1924 the Indian physicist Bose made important theoretical calculations regarding light particles. He sent his results to Einstein who extended the theory to a certain type of atom. Einstein predicted that if a gas of such atoms were cooled to a very low temperature all the atoms would suddenly gather in the lowest possible energy state. The process is similar to when drops of liquid form from a gas, hence the term condensation. Seventy years were to pass before this year's Nobel Laureates, in 1995, succeeded in achieving this extreme state of matter."

The Raman effect was discovered in 1928. Interestingly, between 1923-25 there were 3 independent reports on this phenomenon respectively, by Seshagiri Rao and Raman (1923), by K. R. Ramnath (1923), and by K. S. Krishnan (1925). But all of them initially described this as feeble or very weak fluorescence. In 1923, Compton discovered the inelastic scattering of x-ray photons (Compton effect) and received the Nobel Prize in 1927. Around the end of 1927, it suddenly struck to Raman that their earlier observation may be an optical analogue of the Compton effect. The initial experiments were done using sunlight as a source and eye as a detector. Realizing the great importance of his discovery, Raman wrote a letter to G. D. Birla seeking his financial help to buy a spectrograph and he promised that he would bring a Nobel Prize if he gets the instrument. G. D. Birla and another businessman gave him the money. On Feb 16, 1928 Raman sent the paper to "Nature." He was so anxious that somebody may report the discovery before him that he called a Press Conference on Feb 28, 1928 and announced this discovery. This is the reason why 28 February is considered as the National Science Day in India.

Raman kept his promise. Around the middle of November 1930 he got the telegram that he has been chosen for the Nobel Prize in Physics. The News obviously started great festivities but some serious persons inquired how Raman would get a reservation in the ship to go to Sweden at such a short notice. Raman smilingly said that as early as in July 1930, he booked 2 tickets in advance to go to Sweden- one for himself and the other for his wife.

Though Raman was so confident about getting the Nobel, he created a "scene" at the Nobel Prize ceremony. Tears started flowing out of his eyes as soon as he received the Nobel citation. Later, he pointed out that he could not control his emotions when he saw the British flag on his designated chair which reminded him that he did not get the Prize as a citizen of an Independent country but as a subject nation. Though Raman never took part in politics he displayed his patriotism in his own way. In the official address at the Nobel ceremony he talked about the freedom fighters of India who were languishing in British jails. Through out his life he celebrated 2 October (Gandhiji's birth day) by delivering a Gandhi Memorial lecture. He gave the last Gandhi lecture barely 7 weeks before his death (21 Nov, 1970).

It may be mentioned that Raman was the first Asian to get a Nobel Prize in any branch of Science. The name of Raman is also permanently engraved in modern Physics. Even today Raman Spectroscopy is finding more and more new applications ranging from chemical analysis to early detection of life threatening diseases. For this reason, when the American Chemical Society decided to honour certain discoveries as "International Historic Chemical Landmarks" they identified the Discovery of the Raman Effect as the first one in the whole of Asia. In 1998, a special plaque was presented to IACS by the American Chemical Society as a mark of this honour.

In 1933, Raman was selected as the first Indian Director of the Indian Institute of Science, Bangalore. In the same year Hitler came to power in Germany and because of fascism many scientists left Germany. Raman invited many of them (Schordinger, Born, Goldschmidt and others) to Bangalore. Max Born actually came to Bangalore and spent 3 months. However, a group of professors (mostly European) started a vicious campaign against him. About this Born wrote to Lords Rutherford (who always supported Raman)

> "I do not like to bother you with Indian affairs. But I think I should send you a pamphlet which I received today. It shows the ugly methods used by Raman's opponents."

Born left India. Raman was also forced to resign as a Director in 1936. He continued to be a Professor at IISc. During this period he established the Indian Academy of Sciences and inspired a stream of brilliant students like G. N. Ramachandran and Vikram Sarabhai (1919-71). Raman was instrumental in bringing Homi Bhabha (1909-67) to India. Bhabha and Sarabhai both came from rich families and were educated abroad. But because of the outbreak of the second World War they stayed back in India. Bhabha joined IISc as a Reader. Raman nominated him for the FRS. In 1946, Bhabha established the Tata Institute of Fundamental Research (TIFR) with the donation received from the Tata trust. TIFR served as the cradle of the Indian Nuclear Research program and Bhabha was the First Chairman of the Atomic Energy Commission. Vikram Sarabhai initiated our space research program. At the time of his rather untimely death, Sarabahai was Chairman of both the Indian Space Research Organization and the Atomic Energy Commission.

In the late thirties the scientists of India played a central role in drawing up plans for development in independent India. When Netaji Subhash Bose was the President of the Indian National Congress, the first planning commission was established with Nehru as its chairman. An outcome of this was the establishment of the Bureau of Scientific and Industrial Research (BSIR) in 1940. S. S. Bhatnagar (1894-1955) an eminent physical chemist who worked at Lahore and Benaras Hindu University was chosen as the first secretary of BSIR. In 1942, BSIR was renamed CSIR and Bhatnagar became the first Director-General. Because of the initiatives taken jointly by Nehru and Bhatnagar within 2-3 years after Independence at least half a dozen CSIR institutes were established in India. That India had so many institutions so quickly after Independence was described as the "Nehru-Bhatnagar Effect" by Raman. Bhatnagar was also the first Chairman of UGC (University Grants Commission) when it was established after Independence.

Nehru's close association with the scientists was also reflected in many other things. When the idea of IITs was first mooted, Nehru chose the first IIT to be situated at Hijli Jail in Khargpur. Hijli was infamous for the incident when police fired at unarmed prisoners. Two persons (Santosh Mitra and Tarakeswar Sen) were killed in the firing. One of them (Santosh Mitra) was a brilliant student and first class first in BA in Philosophy in Calcutta University. The idea of Nehru and J. C. Ghosh the first Director of IIT, Kharagpur was to remind the students about the sacrifice of the freedom fighters and the price they paid. In 1956, at the first convocation of IIT, Kharagpur Nehru said,

> "Here, in the place of that Hijli Detention Camp stands this fine monument of India today, representing India's urge, India's future in the making. This picture seems to be symbolic of the changes that are coming to India."

It is evident that the resurgence of science in India was a National Movement. It involved social and political leaders (from Rammohan to Nehru), philanthropists (Tata, Maharaja of Mysore, Palit, Ghosh and others) and institute builders (M. L. Sircar, Asutosh, and others). The prevailing freedom movement motivated the rich and the elite to generously support scientific research. The scientists in their turn, did their best to repay their indebtedness to the sponsors and also to "show the West that we are equal to them." In this respect Raman, Bose, Saha, Chandresekhar, Bhabha- all were a kind of freedom fighters who endeavoured to make our country scientifically and technologically independent.

It will be good for us to remember the following quotation of Raman,

> "I can assert that the quality of the Indian mind is equal to the quality of any Teutonic, Nordic or Anglo-Saxon mind. What we lack is perhaps courage. We need a spirit of victory. If that indomitable spirit were to rise, nothing can hold us from achieving our rightful destiny."

## References

D.P. Chattopadhyay, *"History of Science and Technology in Ancient India,"* Vol. 1 (1986) and Vol. 2 (1991) (Firma KLM, Kolkata).

J.C. Needham, forward of ref. 1 (vol. 1).

P.C. Ray, *"Life and Experiences of a Bengali Chemist,"* Chuckervarty and Chatterjee, Kolkata (1935).

*"A Century,"* Indian Association for The Cultivation of Science (1976).

*"Gleanings of the Past and the Science Movement, (Diary of M.L. Sircar)"* Arun Kumar Biswas, The Asiatic Society, Calcutta (2000).

D.P. Sengupta, *"Sir Asutosh Mukherjee - Educationist, leader and institution builder," Current Science* Vol. 78, p. 1566 (June 25, 2000).

B.P. Radhakrishna, *"Pramatha Nath Bose," Current Science* Vol. 72, p. 222 (February 10, 1997).

V. Mukherjee, *"Builders of Modern India - Jagadish Chandra Bose,"* Publications Divison, Government of India (1983).

D.T. Emerson, *IEEE Transactions on Microwave Theory and Techniques*; Vol. 45, No. 12, pp. 2267-2273 (December 1997).

P.C. Ray, "A History of Hindu Chemistry," Centenary Edition, Shaibya, Kolkata (2002).

B.V. Raman, *"Journey into Light - Life and Science of C.V. Raman,"* Penguin.

In: The Best of Indian Physics
Editor: B. G. Sidharth, pp. 57-92
ISBN 1-59454-115-9


*Chapter 7*

# BOSE STATISTICS: A HISTORICAL PERSPECTIVE

***Partha Ghose***
S. N. Bose National Center for Basic Sciences, Calcutta

## 1 Planck's Derivation of His Law of Black-Body Radiation

In order to appreciate the novelty and importance of Bose's work in its historical perspective, it is necessary first to give a brief account of Planck's original derivation [Planck, 14 December 1900] of the law of black-body radiation,

$$\rho(\nu, T) = \frac{8\pi h\nu^3}{c^3} \frac{1}{e^{h\nu/kT} - 1} \tag{1}$$

which he had discovered empirically a few weeks earlier (7 October 1900). It consisted of three steps.

First, he established the relation

$$\rho_\nu = \frac{8\pi\nu^2}{c^3} U_\nu \tag{2}$$

between the energy density $\rho_\nu$ of incident radiation at temperature $T$, whose frequency lies between $\nu$ and $\nu + d\nu$, and the average energy $U_\nu$ of a resonator of the same frequency $\nu$ at temperature $T$, on the basis of classical electromagnetic theory. Comparing (1) and (2), he found $U_\nu$:

$$U_\nu = \frac{h\nu}{e^{h\nu/kT} - 1}, \tag{3}$$

In the second step he determined the entropy of these oscillators by integrating $TdS = dU$ where $T$ is taken from (3) as a function of $U$ (for fixed $\nu$). He obtained

$$S = k\left[\left(1+\frac{U_\nu}{h\nu}\right)ln\left(1+\frac{U_\nu}{h\nu}\right) - \frac{U_\nu}{h\nu}ln\frac{U_\nu}{h\nu}\right] \quad (4)$$

The third step was the revolutionary one. He introduced two ideas at this stage that he himself considered as "acts of desperation." He assumed that the total energy $U_N = NU_\nu$ of $N$ oscillators (resonators) was made up of finite energy elements $\varepsilon$ such that $U_N = P\varepsilon$ with $P$ a large number. Then he searched in Boltzmann's work for a permutation measure $W_N$ (the total number of complexions or distributions) of $P$ discrete equal energy values $\varepsilon$ among $N$ oscillators that would correspond to the right hand side of (4) when used in $S_N = klnW_N$. He found

$$W_N = \frac{(P+N-1)!}{P!(N-1)!} \quad (5)$$

Using $P/N = U_\nu/\varepsilon, S_N = NS$ and applying Stirling's formula, he obtained

$$S = k\left[\left(1+\frac{U_\nu}{\varepsilon}\right)ln\left(1+\frac{U_\nu}{\varepsilon}\right) - \frac{U_\nu}{\varepsilon}ln\frac{U_\nu}{\varepsilon}\right] \quad (6)$$

Since $S$ is a function of $(U_\nu/\nu)$ only, it follows from (4) and (6) that

$$\varepsilon = h\nu$$

This is how quantum theory was born. Planck had no more justification for using the formula (5) than that it gave him the result that he was after: "... a theoretical interpretation had to be found at any price, however high it might be," he wrote to Robert William Wood in October 1931. [Mehra and Rechenberg, 1982, 1, 1, 50]. Actually, Planck's combinatorial approach differed from Boltzmann's probabilistic method in that Planck associated $W_N$ with $S_N$, the equilibrium entropy, without maximizing it. For Boltzmann $W_N$ was the number of possible complexions corresponding to the macrostate which can be realized by the largest number of complexions. As pointed out by Ehrenfest, a strict adherence to the accepted principles of statistical mechanics would have led Planck to the classical law of Rayleigh and Jeans! [Ehrenfest, 1905].

## 2 Einstein's Light-Quantum Hypothesis

In 1905 Einstein argued on the basis of his analysis of energy fluctuations of radiation obeying Wien's law that such radiation had statistical properties similar to material particles and so must consist of discrete light-quanta of magnitude $(R/N_0)\beta\nu = h\nu$ [Einstein, 1905]. This light-quantum hypothesis was immediately applied by him to give a reasonable explanation of several radiation phenomena, specially the photo-electric effect. Since these conclusions contradicted the classical electromagnetic theory on which Planck attempted to base his theory of heat radiation, Einstein considered Planck's theory "in some way a counterpart" to his own theory. He subjected Planck's derivation to a critical analysis and came to the following conclusion:

> We must therefore regard the following law as the basis of Planck's quantum theory of radiation : the energy of an elementary resonator can only assume values which are integral multiples of $(R/N_0)\beta\nu$; the energy of a resonator changes in jumps by absorption or emission in integral multiples of $(R/N_0)\beta\nu\cdots$ If the energy of a resonator can alter only in jumps, then for the evaluation of the average energy of a resonator in a radiation cavity, the usual [electromagnetic] theory cannot be used, for the latter does not admit any distinctive energy values for a resonator. [Einstein, 1906]

Two years later Lorentz also came to the same conclusion [Lorentz to Wien, 6 June 1908; Mehra and Rechenberg, 1982, 1, 1, 98]. Planck attempted to modify his theory to take into account the quantum of action *without violating any aspect of classical electrodynamics*, but it gradually became clear from his studies and those of others that it was impossible to establish the theory of black-body radiation entirely on the foundation of Maxwell's classical electrodynamics and the statistical mechanics of Maxwell and Boltzmann.

## 3 Debye's Derivation of Planck's Law

In 1910 Debye gave a new derivation of Planck's radiation law in which he sought to avoid the inconsistencies (as viewed from Maxwell's theory) of Planck's earlier derivations as pointed out by Einstein [Debye, 1910]. Instead of using the relation (2) between the radiation density $\rho_\nu$ and the average energy of the oscillator $U_\nu$, Debye calculated "the probability for a given state of radiation and therefore as is well known, the entropy, using the properties of the state alone without employing resonators." In agreement with Rayleigh and Jeans, he calculated the number $N_\nu d\nu$ of elementary states or vibrational modes (Hohlraum oscillators) contained in a volume $V$ and obtained

$$N_\nu d\nu = \frac{8\pi\nu^2 V d\nu}{c^3} \tag{7}$$

Assuming that an amount of energy $h\nu$ gets distributed over each of these vibrations according to an arbitrary distribution function $f_\nu$, he obtained

$$\rho_\nu d\nu = \frac{8\pi h\nu^3}{c^3} f_\nu d\nu \tag{8}$$

Now, "analogously to Planck," Debye assumed that the probability of distributing $N_\nu f_\nu$ quanta of energy $h\nu$ among $N_\nu d\nu$ vibrations was given by

$$\omega_\nu = \frac{(N_\nu d\nu + N_\nu f_\nu d\nu)!}{(N_\nu d\nu)!(N_\nu f_\nu d\nu)!} \tag{9}$$

He then calculated the distribution function $f_\nu$ in the following way. He calculated the maximum of $\omega_\nu$ with the constraint that the energy remained constant. From this he calculated the equilibrium entropy $S_\nu$ to be $kln(\omega_\nu^{max})$. Finally, using the definition of entropy to be $(dS_\nu/d\rho_\nu) = T^{-1}$, he obtained

$$f_\nu = \frac{1}{e^{h\nu/kT} - 1} \tag{10}$$

Together with (8), this yielded Planck's radiation formula (1).

Two important features of Debye's derivation need to be emphasized as background to Bose's derivation that came fourteen years later. First, it became clear from his derivation that Planck's law follows simply from the assumption that the energy transfer from matter to the heat radiation and vice versa is quantized in units of $h\nu$ and that *no knowledge of the properties of material resonators is needed.* In other words, the quantum hypotheses only determined the transfer of energy from one mode of the radiation to another. This was in keeping with Planck's ideas but not Einstein's. Second, he used Planck's definition of the probabilities $\omega_\nu$, without analyzing its significance.

# 4 Indistinguishability of Planck's Quanta

In 1911 Ladislas Natanson [1911] subjected the Planck-Debye combinatorial procedure to a critical analysis and showed that it was equivalent to distributing $P$ indistinguishable energy elements $\varepsilon$ among $N$ "receptacles of energy" distinguished only by the numbers $j$ of quanta in them so that $N_j$ receptacles contain $j$ quanta each, subject to the constraints $\sum_j N_j = N$ and $\sum_j N_j = P$. He found

$$\omega_\nu = \frac{(N_\nu d\nu)!}{p} \tag{11}$$

$$\prod N_j!$$

$$j = 0$$

Contrary to popular belief, therefore, the indistinguishability of the quanta had already been tacitly assumed by Planck and Deye and this was first noticed by Natanson already in 1911.

Ehrenfest and Kamerlingh Onnes [1914] also came to a similar conclusion in 1914. They however claimed that Planck's procedure was equivalent to distributing $P$ indistinguishable energy elements $\varepsilon$ among $N$ distinguishable resonators and arrived at the expression

$$\omega_\nu = \frac{(N-1+P)!}{P!(N-1)!} \tag{12}$$

which is equivalent to the expressions (9) and (11) when $N >> 1$. They then proceeded to clarify the distinction between Planck's energy quanta which were statistically not independent (they called them "non-disjointed quanta") and Einstein's hypothetical light-quanta which were believed to be statistically independent of one another ("disjointed quanta"). They concluded: "Planck's formal device (distribution of $P$ energy-elements $\varepsilon$ among $N$ resonators) cannot be interpreted in the sense of Einstein's light quanta.

## 5 The Introduction of Phase Space Cells by Planck

Planck himself had come to the view on his own that classical statistical mechanics had to be modified in order to yield his radiation formula (1) rather than that of Rayleigh and Jeans. He reported his important conclusion at the Solvay Congress held in Brussels in 1911. According to Gibbs, the probability of finding a single particle in the element $d^3pd^3q$ of the six-dimensional phase space is given by

$$\frac{e^{-E/kT}d^3pd^3q}{\int e^{-E/kT}d^3pd^3q} \tag{13}$$

where $E = E(p,q)$ is the energy of the system. For a classical one-dimensional oscillator, $E = p^2/2m + \beta q^2/2$ and so the average energy of such an oscillator is $kT$. However, if $E$ can take only the discrete values $E_n = n\varepsilon = nh\nu$ with $n = 0, 1, 2\cdot$, then

$$U = \frac{\sum_n E_n e^{-E_n/kT}}{\sum_n e^{-E_n/kT}} = \frac{\varepsilon}{e^{\varepsilon/kT} - 1}, \tag{14}$$

in agreement with (3). It was this straightforward derivation that led Planck for the first time to make the explicit statement that the energy of an oscillator (and not only its average energy) was an integral multiple of $h\nu$. It also led him to interpret the constant $h$ as a finite extension of the elementary area in phase space. The motions of one-dimensional classical oscillators can be described by a family of concentric ellipses of semi-axes $(2E/\beta)^{\frac{1}{2}}$ and $(2mE)^{\frac{1}{2}}$. Since these oscillators can have any energy $E$, the ellipses form a continuum. However, if it is assumed that these ellipses are separated from each other so that the $n$th ellipse encloses an area

$$\int\int dpdq = nh, \tag{15}$$

then the energies of the oscillators must be integral multiples of $h\nu$. Planck therefore drew the revolutionary conclusion that energy quanta were a consequence of the fundamental condition

$$\int_{E}^{E+\varepsilon}\int dpdq = h \tag{16}$$

This implied a revision of the basic tenets of classical mechanics. He declared: "The framework of classical mechanics, even if combined with the Lorentz-Einstein principle of relativity, is obviously too narrow to account for all those physical phenomena which are not directly accessible to our coarse senses... One should therefore confine oneself to the principle that the elementary region of probability $h$ has an ascertainable finite value and avoid all further speculation about the physical significance of this remarkable constant." In other words, all attempts at finding a classical mechanical explanation of $h$ must be abandoned. As we shall see later, to Bose these considerations (with which he was familiar) brought home extremely important lessons that would influence his own seminal contribution to the subject. First, he took serious note of the conclusion that classical electrodynamics was essentially incompatible with Planck's radiation formula. Second, since the quantum states of a particle were not continuously distributed throughout phase space (as in classical dynamics), he concluded that their number could be counted by simply dividing the total volume of phase space by $h^3$, the volume of an elementary cell. Finally, he took seriously Planck's claim that classical statistical mechanics had to be modified in a special way in order to conform to the requirements of quantum theory.

# 6 Spontaneous and Induced Transitions

The next important step was taken by Einstein in 1916. In 1913 Niels Bohr had proposed his model of the atom with stationary states with discrete energy values [Bohr, 1913,b,c]. The transitions between any pair of these states were assumed to be caused by the exchange of energy quanta $h\nu$ between the atom and the radiation field described by *classical electrodynamics*.

Einstein gave an elegant derivation of Planck's radiation formula by considering Bohr atoms in thermal equilibrium with radiation in a cavity but without making any use of classical electrodynamics [Einstein, 1916a]. He used Boltzmann's principle to write the probability $W_n$ for an atom to be in a stationary state with quantum number $n$ in the form

$$W_n = g_n exp\left(-\frac{\varepsilon_n}{kT}\right), \tag{17}$$

$g_n$ being the statistical weight of the state. He then assumed that a stationary state $n$ may pass to a stationary state $m$ of higher energy ($\varepsilon_m > \varepsilon_n$) by absorbing a light quantum of frequency $V_{nm}$, the rate of transition being $B_n^m N_n \rho_v$($N_n$ being the number of atoms in the stationary state $n$ and $B_n^m$ a proportionality constant characterizing the absorption of radiation of frequency $V_{nm}$ in the presence of external radiation of density $\rho_v$). Transitions from a stationary state $m$ of higher energy to a stationary state $n$ of lower energy by the emission of a light-quantum of frequency $V_{mn} = V_{nm}$ could, however, take place in two ways. Atoms can emit this radiation independent of the external field, their number being $A_m^n N_m$($A_m^n$ being the constant of proportionality).

About this Einstein said: "One can hardly think of it in any other way except as a radioactive reaction." It came to be known later as "spontaneous emission." Atoms can also be induced to emit radiation by the external field, the rate being $B_m^n N_m \rho_v$. Einstein called this "negative radiation." In thermal equilibrium one must have $(A_m^n N_m + B_m^n N_m \rho_v) = B_n^m N_n \rho_v$. Using (17) to calculate $N_n/N_m$, one gets

$$A_m^n g_m + \rho_v \left[B_n^m g_n exp\left(\frac{\varepsilon_m - \varepsilon_n}{kT}\right) - B_m^n g_m\right] \tag{18}$$

At this stage Einstein had to take his guidance from classical theory by essentially applying the "correspondence principle" according to which classical theory should be a limiting case of quantum theory. In the classical limit $\rho_v$ goes to infinity as the temperature $T$ goes to infinity for fixed $v$ (Rayleigh-Jeans law, $\rho_v \alpha T$). This gave $B_m^n g_m = B_n^m g_n$. With this condition (18) could be written as

$$\rho_v = \frac{A_m^n / B_m^n}{exp[(\varepsilon_m - \varepsilon_n)/kT] - 1}, \tag{19}$$

Then Einstein used Wien's displacement law to infer that $A_m^n/B_m^n$ must be proportional to $v^3$, and the Rayleigh-Jeans law in the limit of low frequencies to determine the constant of proportionality to be $8\,\pi h/c^3$. Thus he obtained the relation

$$A_m^n = \frac{8\pi h v^3}{c^3} B_m^n \tag{20}$$

and

$$\varepsilon_m - \varepsilon_n = h\nu \tag{21}$$

which is Bohr's frequency condition. These two relations when substituted into (19) gave Planck's law. This is a remarkable derivation that uses an admixture of the light-quantum hypothesis and Bohr's atomic theory which Bohr regarded as being incompatible, Wien's radiation law and therefore implicitly the Maxwell-Boltzmann distribution for the light-quanta, and the correspondence principle which was a heuristic principle without any fundamental basis. Einstein concluded by saying: "I admit freely, of course, that the three hypotheses concerning outgoing and incoming radiation, do not at all become substantiated results by the mere fact that they lead to Planck's radiation formula. But the simplicity of the hypotheses, the generality with which the consideration can be carried through easily, as well as the natural connection of the consideration employed with the limiting case of Planck's linear oscillator (in the sense of classical electrodynamics and mechanics), persuaded me to regard it as very probable that all this constitutes the fundamental outline of the future theoretical derivation." Although the hypotheses concerning the outgoing and incoming radiation turned out to be correct to a large extent (though not entirely so, as we shall see later), the future derivation (given by Bose) did not make use of them at all. In fact, they became the major bone of contention between him and Bose.
By 1916 the light-quantum hypothesis had received a measure of empirical support from Millikan's verification of Einstein's photo-electric equation. Millikan's paper [1916b] appeared in the March 1916 issue of the Physical Review. Einstein quickly realized from his new derivation of Planck's law that atoms seemed to interact with radiation as if single atoms collided with light-quanta of energy $h\nu$ and directed momentum $h\nu/c$. He wrote to his friend, Michael Besso:

> The fundamental thing is that the statistical consideration, which leads to Planck's formula, has become unified and thereby as general as one can imagine, since one has not assumed anything about the nature of the mediating molecules other than the most general quantum idea. From it follows the result (which was not contained in the paper I sent you) that in each energy transfer from radiation to matter the momentum $h\nu/c$ is also transferred to the molecule. Hence we conclude that every such elementary process is a completely directed event. With that the light-quanta must be considered as being substantiated. [Einstein to Besso, 6 September 1916; Mehra and Rechenberg, 1982, 1,2,515]

However, Einstein was not really satisfied. In a subsequent paper he remarked:

> The weakness of the theory lies, on one hand, in the fact that it does not bring us closer to a connection with the undulatory theory, and, on the other hand, in the fact that it leaves the instant and direction of the elementary processes to chance. [Einstein, 1916b, 1917]

## 7 Reactions to the Light-Quantum Hypothesis: Discovery of the Compton Effect

The majority of physicists, in fact, did not take the light-quantum hypothesis seriously, even after Millikan's work. Millikan himself remarked in his 1916 paper on the photo-electric effect: "Yet the semi-corpuscular theory by which Einstein arrived at this equation seems at present to be wholly untenable." The main difficulty lay, as Einstein himself realized, in accounting reasonably for the interference and diffraction phenomena observed with all kinds of radiation in terms of light-quanta. Evidence in favour of the wave theory was overwhelming and mounting. In 1912 Max von Laue, Walter Friedrich and Paul Knipping had settled the issue in favour of X-rays being electromagnetic waves of extremely short wavelengths through the phenomenon of interference of X-rays in crystals [von Laue, Friedrich and Knipping, 1912]. In 1913 Planck, Nernst, Rubens and Warburg who wished to bring Einstein to Berlin, wrote to the Prussian Ministry of Education:

> That he may sometimes have missed the target in his speculations, as, for example, in his theory of light-quanta, cannot really be held against him. For in the most exact of natural sciences every innovation entails risk. [Clark, 1972, 215]

Bohr had his own reasons for not favouring the light-quantum hypothesis though he realized from the beginning that the laws of electrodynamics had to be abandoned for a proper description of quantum phenomena. The hypothesis ran counter to his plans to build atomic theory on the basis of the correspondence principle.

The general acceptance of the light-quantum hypothesis eventually came in 1923 through the work of Debye [1923] and Compton [1923]. Compton realized that the shift in the wavelength of X-rays scattered by atoms that he observed could not be understood in terms of the classical wave theory of scattering but only in terms of elementary processes of scattering of light-quanta by the electrons in atoms, in which both energy and momentum were conserved. Debye also arrived at the same conclusion independently. Arnold Sommerfeld who was then visiting the University of Wisconsin as Carl Schurz Professor of

Physics came to know about Compton's results. On 21 January 1923 he wrote to Bohr: "The most interesting scientific news I have learned in America is a work of Arthur Compton in St. Louis. According to it the wave theory of X-rays would finally have to be given up." [Mehra and Rechenberg, 1,2, 529] Wherever he went, Sommerfeld referred to Compton's work and its importance for the quantum theory. On his return to Germany he continued to advocate Compton's work. Several months later he wrote to Compton : "Your discovery of the change in wavelength of X-rays also occupies the scientific community in Germany most deeply. I met Einstein and Kossel in August, and we mainly discussed your effect." [Sommerfeld to Compton, October 1923; Mehra and Rechenberg, 1982, 1,2, 530]

# 8 Pauli's Derivation of Planck's Law

Wolfgang Pauli came to know of Compton's discovery in the summer of 1923 when he came to Hamburg on several weeks' leave of absence from Copenhagen. He immediately got interested in the problem of thermal equilibrium between radiation and free electrons. Lorentz had reported in the first Solvay Conference that he could not establish the condition of thermal equilibrium for electrons having a Maxwellian velocity distribution [Lorentz's report in Langevin and de Broglie, 1912]. Pauli wished to re-examine the problem by considering reversible collisions between electrons and directed light-quanta, obeying the laws of conservation of energy and momentum. Pauli [1923] was able to show that thermal equilibrium characterized by Planck's law was possible for electrons having a Maxwellian distribution provided the probability of collisions per unit time was given by

$$A\rho_\nu + B\rho_\nu\rho'_\nu, \tag{22}$$

where $A$ and $B$ are the coefficients of Einstein's radiation theory [Einstein, 1917] and $\rho_\nu$ and $\rho'_\nu$ the densities of radiation before and after the collision respectively. Pauli showed that the first term was dominant for radiation satisfying Wien's law while the second term was more important for radiation of longer wavelengths satisfying the Rayleigh-Jeans law. Pauli interpreted this as the effect of interference fluctuations of classical waves. Einstein and Ehrenfest welcomed Pauli's work and clarified and generalized it to cases in which the elementary processes involved more than two light-quanta [Einstein and Ehrenfest, 1923]. In particular, they discussed the "paradoxical" second term in (22) "which implies that the number of the elementary processes of scattering occurring per unit time at an electron (which is at quasi rest) increases faster than being proportional to the radiation density $\rho'$ of frequencies $\nu'$, which the radiation quantum possesses after being modified by the elementary process." They concluded that this term followed from a detailed consideration of all interaction processes between radiation and electrons, including the induced emission of light (Einstein's "negative radiation").

# 9 Bose's Derivation

While the leading European and American physicists were skeptical, even hostile to Einstein's light-quantum hypothesis, two unknown young physicists in India, M.N. Saha and S.N. Bose grasped its importance and started using it with great success. Within two years of Einstein's 1917 paper from which he drew the revolutionary conclusion that a light-quantum carried directed momentum $h\nu/c$, M.N. Saha, a close friend of Bose, applied it to explain the existence of radiation pressure on objects of molecular size [Saha, 1919]. On the basis of the classical theory Nicholson [1914] and Klotz [1918] had shown that the radiation pressure on particles of molecular size should become totally "evanescent."

Saha applied "quantum theory in the place of the old continuous theory of light. Instead of assuming that "light" is spread continuously over all points of space," he supposed that "they are localized in pulses of energy $h\nu \cdots$ Let this pulse encounter a molecule $m$ and be absorbed by it. Then in doing so the molecule will be thrust forward with an impulsive momentum of $h\nu/c^2$ and the momentum $h\nu/c$; the absorption of the pulse by the molecule may be taken as a case of inelastic impact, whole momentum being communicated to the molecule. The velocity with which the molecule will move forward $= h\nu/mc$." Saha concluded that "radiation-pressure may exert an effect on the atoms and molecules which are out of all proportion to their actual sizes." This must have been the first application of Einstein's hypothesis that light-quanta carried a directed momentum $h\nu/c$.

In March 1924 Saha visited Dhaka and stayed with Bose. While teaching postgraduate students Bose had keenly felt the need for a derivation of Planck's law free of logical difficulties. Saha drew Bose's attention to the papers of Pauli [1923] and their connection to Einstein and Ehrenfest [1923] and their connection to Einstein's paper of 1917. In an interview with J. Mehra [Mehra and Rechenberg, 1,2,565] Bos4e recalled: "What seemed to be happening in Pauli's work was that in order to apply the quantum conditions you had to know exactly what was going to happen afterwards."

Stimulated by his discussions with Saha, Bose began to study carefully the works of Planck by his discussions with Saha, Bose began to study carefully the works of Planck [1900,1910], Peter Debye [1910,1923]. This resulted in two papers that he wrote in June 1924 [Bose, 1924a, 1924b] and sent to Einstein for his views and for publication in Zeitschrift fur Physik. Einstein translated both of them into German and had them published in Zeitschrift fur Physik in 1924. To both of them he added his own comments.

It is clear from these two papers of Bose which he always regarded as integral parts of each other that he clearly recognized the following points:

(1) The fundamental assumptions of the quantum theory are incompatible with the laws of classical electrodynamics [Bose, 1924a].
(2) All previous derivations had made use of relation (2) "between the density of radiation and the average energy of an oscillator, and they make assumptions about the number of degrees of freedom of ether, which enters the above equation (the first factor on the right hand side). This factor could however be derived only from the classical theory. This is an unsatisfactory feature in all derivations and it is not surprising that efforts are made again and again to give a derivation free from this logical flaw." [Bose, 1924a]
(3) Even Einstein's 1916/1917 derivation, although "remarkably elegant," makes use of "Wien's displacement law and Bohr's correspondence principle. Wien's law is based on classical theory and the correspondence principle assumes that the quantum theory agrees with the classical theory in certain limiting cases." [Bose, 1924a].
(4) "In all cases it appears to me," wrote Bose [1924a] "that the derivations are not sufficiently justified from a logical point of view. On the other hand, the light-quantum hypothesis combined with statistical mechanics (as adapted by Planck to conform to the requirements of quantum theory) appears sufficient for the deduction of the law independent of classical theory" (our italics). In other words, the incompatibility between Einstein's light-quanta ("disjointed quanta") and Planck's law (implying "undisjointed quanta") can be removed by adapting the statistical mechanics of light-quanta to conform to the requirements of quantum theory, as proposed by Planck.
(5) "The problem of thermodynamic equilibrium of radiation in the presence of material particles can however be studied using the methods of statistical mechanics, independently of any special assumption about the mechanism of the elementary processes on which the energy exchange depends." [Bose, 1924b].

Bose therefore proceeded in four steps.

First step: He took Einstein's light-quantum hypothesis seriously and treated black body radiation as a collection of light-quanta enclosed in a volume $V$. If there are $N_s$ light-quanta of energy $h\nu_s(s = 0 \quad s = \infty|$, the total energy $E$ is

$$E = \sum_s N_s h\nu_s = V \int \rho_\nu d\nu \tag{23}$$

$\rho_\nu$ can be determined from this simple relation if $N_s$ is known. So, the problem is to determine $N_s$. "If we can give the probability for each distribution characterized by arbitrary values of $N_s$, then the solution is determined by the condition that this probability is a maximum subject to the subsidiary condition" (23) [Bose,1924a]. Bose then proceeded to find this probability.

Second step: Before he could do that, he needed to compute the number of states of a light-quantum whose frequency lies between $\nu_s$ and $\nu_s + d\nu_s$. In order to be able to do this, he extended Planck's idea of dividing up the phase space of a material oscillator into elementary cells of volume $h^3$ to the phase space of light-quanta. In other words, he extended Planck's "first quantization" of material oscillators to the radiation field itself. This was the most novel and crucial input.

Emboldened by Compton's discovery, he assumed the momentum of a light-quantum of frequency $\nu$ to be $h\nu/c$ in the direction of its motion. He must have also been aware of Saha's success in explaining radiation pressure on molecular-size objects with the help of this hypothesis [Saha,1919]. The instantaneous state of such a light-quantum, he stated, is characterized by a point in six-dimensional phase space which is constrained to lie on the surface of a cylinder defined by

$$p_x^2 + p_y^2 + p_z^2 = \frac{h^2\nu_s^2}{c^2} \tag{24}$$

Then the phase space belonging to the frequency interval $d\nu_s$ is

$$\int dxdydzdp_x dp_y dp_z = 4\pi V \left(\frac{h\nu_s}{c}\right)^2 \frac{hd\nu_s}{c} = 4\pi \frac{h^3\nu_s^2}{c^3} V d\nu_s \tag{25}$$

Dividing this by $h^3$, the "ultimate elementary region" [Bose to Einstein, 4 June 1924], Bose obtained $4\pi V \nu_s^2 d\nu_s/c^3$. Then the following statement occurs: "In order to take into account the fact of polarization, it seems however appropriate to multiply this number once again by 2′ (our italics) to obtain $8\pi V \nu_s^2 d\nu_s/c^3$ as the number of cells belonging to $d\nu_s$. No explanation is offered as to how this "fact of polarization," an essentially classical concept, can be understood in terms of the light-quantum hypothesis, although Bose claimed to deduce this factor independent of the classical electrodynamics [Bose to Einstein, 4 June 1924].

Bose had always maintained privately that he did offer a quantum theoretic explanation but Einstein removed it from his translation and substituted it by the statement about the polarization factor. In his letter to Bose, Einstein simply wrote: "You are the first person to derive the factor quantum theoretically, even though because of the polarization factor 2, not wholly rigorously" [Einstein to Bose, 2 July 1924].

What was the explanation that Bose claimed to have offered? It was that light-quanta carried an intrinsic spin that could take only the values $\pm h/2\pi$. There is only one recorded evidence of this because Bose's original manuscript in English is missing from the Einstein archives. It appears in a paper by C.V. Raman and S. Bhagavantam [1931]. The paper is entitled "Experimental Proof of the Photon Spin." They write:

> In his well-known derivation of the Planck radiation formula from quantum statistics, Prof. S.N. Bose obtained an expression for the number of cells in phase space occupied by the radiation, and found himself obliged to multiply it by a numerical factor 2 in order to derive from it the correct number of possible arrangements of the quantum in unit volume. The paper as published did not contain a detailed discussion of the necessity for the introduction of this factor, but we understand from a personal communication by Prof. Bose that he envisaged the possibility of the quantum possessing besides energy $h\nu$ and momentum $h\nu/c$ also an intrinsic spin angular momentum $\pm h/2\pi$ round an axis parallel to the direction of its motion. The weight factor 2 thus arises from the possibility of the spin of the quantum being right-handed or left-handed, corresponding to the two alternative signs of the angular momentum. There is a fundamental difference between this idea, and the well-known result of classical electrodynamics to which attention was drawn by Poynting and more fully developed by Abraham that a beam of light may in certain circumstances possess angular momentum ... Thus, according to the classical field theory, the angular momentum associated with a quantum of energy is not uniquely defined, while according to the view we are concerned with in the present paper, the photon has always an angular momentum having a definite numerical value of a Bohr unit with one or other of the two possible alternative signs.

Why did Einstein remove Bose's explanation? Probably he felt it was too radical and unnecessary at that stage. However, instead of removing it altogether, he could have added a suitable comment to the one he did make at the end of the paper. In fact, he did not hesitate to add a dissenting note to Bose's second paper [1924b] as we shall see later.

Be that as it may, let us proceed with Bose's derivation. Having divided the phase space volume belonging to the frequency range $\nu_s$ and $\nu_s + d\nu_s$ by $h^3$, he wrote: "the total number of cells must be regarded as the number of possible arrangements of a quantum in a given volume" (our italics). Classically there are infinitely many ways of arranging or distributing a particle within a phase space cell of finite extent corresponding to the infinitely many points contained within it. However, according to Planck, as we have seen, quantum states are not continuously distributed in phase space but are separated from one another by finite amounts determined by $h$. Thus, an elementary cell of volume $h^3$ can contain at most one quantum state, and the problem of counting the number of quantum states reduces to the counting of these elementary cells. This is a totally new interpretation of the factor $A_s = 8\pi\nu_s^2 V d\nu_s / c^3$. (In classical wave theory it is the number of transverse standing modes in the interval $d\nu_s$ in a volume $V$). In his covering letter to Einstein, Bose wrote:

> Respected Sir,
>
> I have ventured to send you the accompanying article for your perusal and opinion. I am anxious to know what you think. You will see that I have tried to deduce the coefficient in Planck's law independent of the classical electrodynamics, only assuming that the ultimate elementary regions in Phase-space have the content $h^3$. [Bose to Einstein, 4 June 1924]

On 12 July 1924 Einstein wrote to Ehrenfest: "The Indian Bose has given a beautiful derivation of Planck's law, including the constant [i.e. $8\pi\nu^2/c^3$]."

Third Step: Having demonstrated that the factor $8\pi\nu^2/c^3$ in Planck's law corresponds to the total number of quantum states of radiation, Bose then proceeded "to calculate the thermodynamic probability (macroscopically defined) of a state" in the sense of Planck (our italics). This is equal to the number of different microscopically defined states by means of which the macroscopic state can be realized. If all the cells (states) were considered to be distinct from one another, the answer would be $A - s!$ However, this is not the case. Since each cell is to be counted as a single quantum state, if follows that the permutations of the light-quanta within a cell do not produce new states. Consequently, the cells can be partitioned into distinct classes characterized solely by their occupation numbers. Let there be a particular distribution of $N_s$ quanta among $A_s$ states such that there are $p_1^s$ states with one quantum, $p_2^s$ states with two quanta, and so on. Then, following Boltzmann and the usual procedure of statistical mechanics for these states (not the quanta), one can immediately write down the possible number of such distributions as

$$W_s = \frac{A_s!}{p_0^s!p_1^s!p_2^s!\cdot} \tag{26}$$

which is precisely the expression (11) given by Natanson. The "thermodynamic probability" of "the state defined by all $p_r^s$" is then clearly $W = \prod_s W_s$ which is equivalent to the Debye-Planck expression (9).

Fourth Step: Bose then followed the standard procedure, already followed by Debye [1910], of maximizing $lnW$ subject to the constraints $E = \sum_s N_s h\nu_s$ and $N_s = \sum_r rp_r^s$ to obtain Planck's law. Notice that the total number of photons $N = \sum_s N_s$ and there is no constraint on $N$. It turns out that the constraint on $N_s$ is not necessary, but Bose did not notice this.

Notice that the only departure of the expression (26) from the usual Boltzmann expression lies not in the product of factorials in the denominator but in putting the multiplicative factor $(g_i)^r$ corresponding to the possible arrangements of $r$ quanta within each cell $i$ having $g_i$ levels equal to unity, as demanded by the division of phase space into elementary cells of volume $h^3$. (In his second paper [1924b] written more or less at the same time as the paper under discussion and received by Einstein only five days later, Bose used these factors, as we shall see later, for material particles obeying Maxwell-Boltzmann statistics.) This is precisely the adaptation of statistical mechanics by Planck "to conform to the requirements of the quantum theory" that Bose referred to in the third paragraph of his paper.

Was Bose aware of the analysis of the Planck-Debye definition of the probability by Natanson and by Ehrenfest and Kamerlingh Onnes? One cannot tell. He certainly did not refer to their papers. It is reasonable, however, to assume that he was not aware of the later work because if he had been, he would have explained how their conclusion, namely that "Planck's formal device (distribution of $P$ energy elements $\varepsilon$ among $N$ resonators) cannot be interpreted in the sense of Einstein's light-quanta" [Ehrenfest and Kamerlingh Onnes, 1914, p.873], could be reconciled with his own starting point, namely the combination of the light-quantum hypothesis with Planck's counting device. The explanation, of course, lies in the fact that Planck's law reduces to Wien's law in the limit $h\nu/kT >> 1$, and that it is only in this limit that the light-quanta have Maxwell-Boltzmann behavior as inferred by Einstein. It is one of those quirks of history that Einstein should have been able to infer the quantum nature of radiation from its approximate classical statistical behaviour in a certain domain and thus missed the true nature of its statistics. It was left to Bose to discover it, possibly uninhibited by the analysis of Ehrenfest and Kamerlingh Onnes and their conclusion. Hence Einstein's enthusiastic welcome of Bose's work as an important and new support to his favourite light-quantum hypothesis, making it compatible with Planck's law.

## 10 The Birth of a New Statistics: Einstein Applies Bose's Method to Ideal Gases

Bose's counting of states, of course, implied a new statistics for the light-quanta. Neither he nor Einstein seemed to have quite grasped the importance of this initially. In his letter of 2 July 1924 to Bose, Einstein compliments him for being the first to derive the factor $A_s$ quantum theoretically but does not write a word about the new statistics. "In my opinion," wrote Einstein in a footnote to Bose's paper, "Bose's derivation signifies an important advance. The method used here gives the quantum theory of an ideal gas as I will work out elsewhere."

He extended Bose's method to material particles in three communications to the Prussian Academy in Berlin on 10 July 1924, 8 January 1925 and 29 January 1925 [Einstein, 1924, 1925a, 1925b] without informing Bose or asking for his collaboration. Bose came to know of the comment in the footnote shortly before he left for Europe and of the first of communications to the Prussian Academy on his arrival in Europe in October 1924. In the first paper this is how Einstein described Bose's method:

> [First]: The phase space of an elementary object (here of a monatomic molecule and in Bose's case of a light-quantum), associated with a given (three-dimensional) volume, is divided into "cells" of extension $h^3$. [Second]: If many elementary objects are present then their (microscopic) distribution, which plays a role in thermodynamics, is determined by the particular manner how the elementary objects are distributed among the cells. [Third]: The "probability" of a macroscopically defined state (in the sense of Planck) is equal to the number of different microscopic states by means of which the macroscopic state can be thought to be realized. [Fourth]: The entropy of the macroscopic state and, therefore, the statistical and thermodynamic behaviour of the system is determined by Boltzmann's theorem relating entropy and probability." [Einstein, 1924, p.261]

Again, there is no mention of the new statistics! In applying Bose's method to ordinary material atoms, Einstein had to introduce an additional parameter (the chemical potential which vanishes for light-quanta) to take into account the conservation of their number. He drew a very important conclusion from this:

> According to the theory presented here, Nernst's theorem is satisfied in the case of ideal gases. To be sure, our formulae cannot immediately be applied to extremely low temperatures, for we have assumed in their derivation that the $p_r^s$ change only relatively infinitely little if $s$ is altered by 1. Still one recognizes at once that the entropy must vanish at the absolute zero of temperature. The reason is that then all molecules are in the first cell; and for this state there exists only one distribution of molecules according to our counting method. Hence our assertion is immediately proved to be correct. [Einstein, 1924, p.265] (our italics)

This is a conclusion of far-reaching consequence, yet there is no mention of any new statistics for the molecules; the emphasis is on Bose's counting method for states (one quantum state per cell). It was only six months later on 8 January 1925 that Einstein discussed the difference between the counting methods of Boltzmann and Bose and recognized the fact, already pointed out by Ehrenfest [1924], that the new statistics for molecules was inconsistent with their statistical independence.

However, while Planck [1925] considered the analogy with radiation a weak point in Einstein's gas theory, the statistics of radiation and that of material particles being completely different in his opinion, Einstein regarded this result differently. He concluded: "This result represents in itself a support of the view concerning the deep natural relation between radiation and gas, since the same statistical treatment which leads to Planck's formula establishes - when applied to ideal gases - the agreement with Nernst's theorem." [Einstein, 1925a, p.7] (our italics) He was of the opinion that an understanding of the "mysterious influence" among the molecules would come in the future.

## 11 The Birth of Wave Mechanics

Based on the new counting method, he showed in his second paper communicated on 8 January 1925 that the mean square energy fluctuation of the gas molecules is given by an expression which is the sum of two terms, one corresponding to the Maxwell-Boltzmann statistics of non-interacting molecules and the other to interference fluctuations associated with wave phenomena. (There exists in the Einstein Archives in the Jewish National and University Library, Jerusalem, a two page calculation done by Bose entitled "Fluctuation in density" which ends with this fundamental result. The document is undated and exists in the Scientific Correspondence File Folder "B-Misc.-II." Bose was in Paris from 18 October 1924 until the first part of October 1925). From this he immediately drew attention to the importance of Louis de Broglie's Doctoral thesis (which he heard of from Paul Langevin and asked for a copy which he received and read in December 1924 [Jammer, 1966, 249]) in which de Broglie had attached wave properties to ponderable matter in analogy with the wave-particle duality of radiation. Einstein wrote: "I shall discuss this interpretation in greater detail because I believe that it involves more than a mere analogy." He then went on to explore various consequences of de Broglie's hypothesis.

When Erwin Schrodinger whose principal research during his early years in Zurich (1921-27) was on the statistical thermodynamics of ideal gases, read Einstein's first paper [1924a] on the quantum theory of an ideal gas, he also did not understand that the molecules were being treated as being indistinguishable. On 5 February 1925 he wrote to Einstein suggesting a possible error in his probability formula. Einstein replied explaining that the "quanta or molecules are not treated as independent of one another" and even gave a little diagram to explain the difference between Bose's counting method and that of Boltzmann. He concluded his letter with the emphatic statement: "There is certainly no error in my calculation." [Walter Moore, 1989, 193] It was Einstein's remark in his second paper that de Broglie's idea "involves more than a mere analogy" that really induced Schrodinger to study the "de Broglie-Einstein undulatory theory" as he called it at that time, and eventually led to his discovery of wave mechanics in 1926 [Schrodinger, 1926a, 1926b and 1926c]. Schrodinger later said that "wave mechanics was born in statistics." (our italics)[Walter Moore, 1989, 188] He also wrote to Einstein on 23 April 1926: "By the way, the whole thing would not have started at present or at any other time (I mean, as far as I am concerned) had not your second paper on the degenerate gas directed my attention to the importance of de Broglie's ideas". [Jammer, 1966,257]

## 12 Further Developments

Meanwhile in July 1925 Werner Heisenberg had sent a paper to Zeitschrift fur Physik in which he gave a preliminary account of matrix mechanics [W Heisenberg, 1925]. Schrodinger proved the complete equivalence of his wave mechanics and the Heisenberg-Born-Jordan matrix mechanics [Schrodinger, 1926d].

A whole train of rapid developments occurred. Pauli discovered his Exclusion Principle within six months of Bose's paper [Pauli, 1925]; Fermi published his paper on the statistics of particles obeying Pauli's principle in early 1926 [Fermi, 1926]; Dirac linked the Bose and Fermi statistics of particles to the symmetry properties of their wave functions and named them "bosons" and "fermions" [Dirac, 1926]; in 1927 Dirac invented "second quantization" of the electromagnetic field [Dirac, 1927] which was soon followed by a similar method for Fermi fields developed by Jordan and Wigner [1928]; later Pauli [1940] proved the spin-statistics theorem in relativistic quantum field theory; Pauli[1927] also applied Fermi-Dirac statistics to the paramagnetism of an electron gas and Sommerfeld [1927,1928] applied it to the electrons in a metal; and Fritz London [1938] suggested that the superfluidity of liquid helium ($^4He$) was related to Einstein-Bose condensation. Thus the developments of Schrodinger's wave mechanics and that of quantum statistical mechanics were directly triggered off by Bose's first paper followed by Einstein's far-sighted extension of his method to ideal gases of material particles and his championing of de Broglie's ideas which soon received experimental confirmation in the hands of Davisson and Germer [1927] and G.P. Thomson [1927].

Years later this is how Schrodinger summed up the position:

> The two equivalent ways of looking at (7.19) either as counting the number of quantum states of a particle, or as counting the number of wave-mechanical proper vibrations of the enclosure, interest us for this reason. The second attitude makes us think of the '$n_s$ particles present in state $\alpha_s$' as of a proper vibration (or a "hohlraum" oscillator to use a customary phrase) in its $n_s$th quantum level. (This attitude really corresponds to so-called second quantization or field quantization.)$n_s$ becomes a quantum number and the stipulation that the system of quantum numbers
>
> $$n_1, n_2, n_3, \cdots, n_s \cdots,$$
>
> determines only one state of the gas, not a class of
>
> $$\frac{n!}{n_1!n_2!\cdots n_s!\cdots}$$
>
> states, ceases to be a strange new adoption, and comes into line with the ordinary view about quantum states and their statistical weight (viz. equal for any two of them).
>
> It is the first, the particle attitude, that has suggested the term "new statistics" which is frequently used. And that is why this idea of new statistics did not, originally, arise in connection with heat radiation, because here the wave point of view was the historical one, the classical one–nobody thought of any other at the outset. The

> wave picture was considered to be (and historically was) the classical description. The quantization of the waves therefore duly appeared to be a "first" quantization and nobody thought of anything like "second quantization."
>
> Not until the idea of photons had gained considerable ground did Bose (about 1924) point out that we could, alternatively to the "hohlraum" oscillator statistics, speak of photon statistics, but then we had to make it "Bose statistics." Very soon after, Einstein applied the same to the particles of an ideal gas. And thereupon I pointed out that we could also in this case speak of ordinary statistics, applied to the wave mechanical proper vibrations which correspond to the motion of the particles of the gas [Schrodinger 1952, 49-50].

We wish only to point out, as explained above, that Bose did not actually "speak of photon statistics"–it was implied by his counting of quantized photon states, and this only became gradually clear with Einstein's application of Bose's counting method to ideal gases, although Ehrenfest and Natanson had already pointed this out in 1911 and 1914 respectively in relation to Planck's counting method of December 1900 which Bose showed was applicable to light-quanta.

## 13 Did Bose Make a "Shot in the Dark?"

Some recent widely read scientific biographies have given rise to the impression that Bose made "a shot in the dark" with three assumptions all of which turned out to be correct but none of which he tried to justify "by arguments of any kind; they seemed to appear intuitively to his mind, perhaps because his thoughts were uncluttered by the ongoing controversies of the European physicists." [W. Moore, 1989, 181; A. Pais, 1982, 428]

The assumptions were that photons were (a)massless particles capable of existing in two states of polarization, (b) indistinguishable and (c) their number was not necessarily conserved.

The record must be put right. That photons are massless particles with energy $h\nu$ and directed momentum $h\nu/c$ was proposed by Einstein and only used by Bose. That they exist in two possible "polarization states" was inferred by Bose from phase space considerations which yield for the number of quantum states between $\nu_s$ and $\nu_s + d\nu_s$ the factor $4\pi V \nu_s^2 d\nu_s/c^3$ rather than $8\pi V \nu_s^2 d\nu_s/c^3$, the factor that occurs in Planck's formula. Actually, as we have already discussed, in order to account for this factor of 2, Bose had proposed that photons carried one unit of intrinsic spin which could take the values $\pm 1$. Whoever had heard of particles with polarization? Moreover, "polarization" was a wave concept which Bose wanted to avoid. However, Einstein quietly dropped it and the statement regarding the polarization factor was most probably inserted by him. Since Bose's original manuscript in English is missing from the Einstein archives and Bose did not keep a copy, there is no way to verify this.

How about "indistinguishability?" Again, as we have seen, although Bose did not explicitly use the word "indistinguishable" which is by itself rather vague, he id use Planck's division of phase space into elementary cells of volume $h^3$ to justify his counting method for states (which implied indistinguishability of the quanta), namely that there cannot be more than one quantum state in an elementary cell. "In any case," he wrote, "the total number of cells must be regarded as the number of possible arrangements of a quantum in the given volume." What is not usually appreciated in usual expositions of Bose statistics is that Bose did not count the number of ways of distributing individual photons over a set of states. Instead, he calculated the thermodynamic probability of a macroscopic state by counting the number of ways in which the macroscopic state can be thought to be realized through given distributions of the number of photons over the micro-states (phase space cells).

Finally, what about non-conservation of photons? This is a non-issue. Since, like Debye [1910], Bose considered the statistical thermodynamics of an open system of radiation in thermal equilibrium and did not include the molecules of matter with which the radiation constantly exchanged energy (which he did in his second paper which we shall turn to presently), there was no question of conservation of energy quanta even if treated as massless particles. In his 1917 derivation of Planck's law Einstein did not impose the conservation of light-quanta which he certainly regarded as massless particles interacting with molecules. Since Bose was not doing anything new or different in this respect, there was no original or deep comment he could have made on this issue. Nevertheless, it is true that Bose, like most others who have made revolutionary contributions, did not, and count not be expected to, understand the full implications of their discoveries.

## 14 Bose's Second Paper(1924)

We now turn to his second paper [1924] on *"Thermal Equilibrium in Radiation Field in the Presence of Matter"*, also translated into German probably by Einstein himself. This paper, completed on 14 June 1924, i.e. more or less at the same time as the first paper, consists of two parts. In the first part Bose derived general conditions for statistical equilibrium of a system consisting of matter and radiation, *independent of any special assumptions about the mechanism of the elementary radiative processes.* In the second part he proposed a new expression for the probability of these elementary radiative processes which differed from that of Einstein. Einstein considered this hypothesis not to be "applicable" to elementary radiative processes and gave two reasons for it in a footnote to the paper. As a consequence, the paper has been largely ignored. However, Bose remained firmly of the opinion for the rest of his life that Einstein had not done him justice.

Bose started off by giving a quick critical review of the most important derivations of Planck's law until then given by Debye [1910], Einstein [1917], Pauli [1923], and Einstein and Ehrenfest [1923] on the scattering of radiation by electrons. Bose considered Debye's 1910 derivation not to be "completely independent of classical electrodynamics" since he derived the factor $8\pi\nu_s^2 V d\nu_s/c^3$ from the normal modes of the ether. He considered the derivations of Einstein [1917], Pauli [1923] and Einstein and Ehrenfest [1923] to be dependent on ad hoc special assumptions about the probability of elementary radiative processes. "The problem of thermodynamic equilibrium of radiation in the presence of material particles," he wrote, "can however be studied using the methods of statistical mechanics, independently of any special assumption about the mechanisms of the elementary processes on which the energy exchange depends. In this way we obtain a general relation which is valid for all special assumptions about the elementary processes and their possibilities." This is in keeping with Kirschhoff's law.

He then went on to say: "If it is possible to write down the thermodynamic probability for any particular distribution of quanta of radiation and the probability for any arbitrary energy distribution among the particles, then the thermodynamic probability for the bigger system consisting of matter and radiation is simply the product of the two probabilities. The condition of equilibrium is simply that the resulting probability must be a maximum." For the thermodynamic probability for radiation, Bose used the expression

$$W = \prod_{\nu} \frac{(A_\nu + N_\nu d\nu)!}{A_\nu! N_\nu d\nu!} \tag{27}$$

where $A_\nu = 8\pi V \nu^2 d\nu/c^3$ which, he stated, had been derived earlier. As references he quoted his own paper and that of Debye [Debye 1910]. (It is curious that the reference to his previous paper is given as "The author, to appear in Phil Mag." This is because Bose had submitted his paper first to the Philosophical Magazine. Not hearing from them for a while, he sent a copy to Einstein. While translating Bose's second paper, the translator obviously forgot to change the reference of the earlier paper to Zeitschrift fur Physik 26, 178-181, 1924). Although expression (27) is indeed the Ansatz used by Debye, it does not appear in Bose's first paper! Instead, the corresponding expression that appears in his first paper is (26), the form given by Natanson. As we have seen earlier, the two are mathematically equivalent, but Bose did not care to explain this, probably because this result must have been fairly well known in the literature in those days.

For the thermodynamic probability for material particles. Bose writes:

> This can be found easily. We want to generalize our assumptions a little so that the case of the Bohr atom with discrete energy levels as well as the case of translational energy of particles can be included. Let the phase space be divided into cells. For every cell there is a probability $g$ that a particle occupies it. The $g$'s are in general equal except for the case of Bohr's atoms. The thermodynamic probability for an arbitrary distribution $n_1, n_2$ etc among the different cells is,

$$\frac{N! g_1^{n_1} g_2^{n_2} \cdots}{n_1! n_2! \cdots} \tag{28}$$

(our italics). Unlike in the case of radiation (expression (26), Bose now uses the factor $\prod g_i^{n_i}$ for material particles and obtains the classical Maxwell-Boltzmann distribution. In other words, he treats material particles as in classical (distinguishable) statistical mechanics but light-quanta according to quantum (indistinguishable) statistical mechanics.

Finally, he writes down the thermodynamic probability for the total system as the product of (27) and (28):

$$W = \prod_s \frac{(A_s + N_s)!}{A_s! N_s!} \prod_t \frac{g_T^{nT} N!}{n_T!} \tag{29}$$

with

$$\sum N_s h\nu_s + \sum n_T E_r = E \tag{30}$$

and

$$\sum N_T = N \tag{31}$$

He then considers an elementary process in which a particle passes from the $r$th cell to the $s$th cell while a light-quantum of frequency $\nu$ changes into a light-quantum of frequency $\nu$. The stationarity of $W$ subject to the conditions (30) and (31) gives

$$\frac{n_r}{g_r}\prod_{\nu}\frac{N_\nu}{N_\nu - A_\nu} = \frac{n_s}{g_s}\prod_{\nu'}\frac{N'_\nu}{N'_\nu + A'_\nu} \tag{32}$$

where

$$\sum h\nu' - \sum h\nu + E_s - E_r = 0 \tag{33}$$

Bose then demonstrated that (32) indeed generalized the results of Pauli [1923] and Einstein and Ehrenfest [1923] without the need for any ad hoc assumptions regarding the elementary radiative processes designed to give Planck's formula. In particular, in the case of Bohr's atoms considered by Einstein in 1917, equation (32) reduces to

$$\frac{n_r}{g_r}\frac{N_\nu}{N_\nu + A_\nu} = \frac{n_s}{g_s} \tag{34}$$

which is Einstein's condition derived on the assumption that atoms in higher energy levels make transitions to lower energy levels in two ways:

1. A spontaneous transition as in radioactivity independent of the state of the external radiation field.

2. An induced transition whose probability depends on the state of the external radiation field.

Transitions from lower to higher energy levels always take place by induced absorption whose probability depends on the state of the external radiation field. Further, Einstein had to postulate certain relations (namely, the opposite limiting Wien and Rayleigh-Jeans laws) among these transition probabilities to obtain Planck's law. Bose's derivation is independent of all these assumptions. It is for this reason that Einstein took the trouble to have the paper translated into German and communicated it to Zeitschrift fur Physik. However, his disagreement with Bose started from the next part of the paper in which Bose proposed a new expression for the probability of an interaction between a particle and a quantum of radiation.

Bose's starting point was the following observation which he considered to be fundamental:

> ..even in a collision no interaction is as probable as the occurrence of any special interaction... From the classical theory... one would expect that there is some interaction whenever a quantum and a

> material particle come together. Therefore here it is a question of a departure from classical theory.
> The assumption made above, that even in a collision no interaction need occur, is quite analogous to the assumption of the stability of stationary states which is so fundamental to Bohr's theory of line emission and can be traced back to the same origin–the probability of existence of the stationary states of the particles. It is interesting to remark in this connection that in the analogous case of a collision of an electron with an atom, experiments show that the electron goes through the atom without changing either the interatomic or its own motion. (our italics)

Bose was obviously referring to the Ramsauer-Townsend effect. He saw a simple way of realizing the non-classical possibility in Planck's cellularization of phase space. Let $p_r$ be the number of cells with $r$ quanta. "In order that the radiation and particle may interact, it must be in a cell occupied by a quantum. The particular cell which the molecule under consideration occupies will pass through all possible states if we observe it long enough, that is, sometimes it is empty, sometimes occupied by 1 quantum, sometimes by 2 quanta etc. The duration of these events will be finally proportional to the equilibrium values of $p_0, p_1$ etc. When $r$ quanta and a particle are together, then either an exchange of energy takes place or nothing happens." (our italics) Therefore $r+1$ different events are possible, namely the exchange of 1 quantum of energy, 2 quanta etc or no exchange at all. Therefore the total number of possible events is

$$p_0 + 2p_1 + 3p_2 + \cdots = A_v + N_v dv_s = \sum + r(r+1)p_r \tag{35}$$

The number of cases in which interaction or energy exchange occurs is

$$p_1 + 2p_1 + 3p_3 + \cdots = N_s dv_s = \sum_r r p_r \tag{36}$$

Consequently the probability of an interaction is

$$P = \frac{\sum r p_r}{\sum (r+1) p_r} = \frac{N_s dv_s}{A_s + N_s dv_s} \tag{37}$$

This is Bose's second fundamental result.

Having derived this result, Bose then considered Einstein's 1917 problem, namely Bohr atoms exchanging energy with the radiation field. In order to be consistent with the fundamental equilibrium condition (34), he found it was sufficient to consider (a) transitions from lower to higher states through absorption of quanta of characteristic frequency $\nu$ and (b) transitions from higher to lower states to be spontaneous, i.e. independent of the external radiation field. The additional assumption of induced emission (negative radiation) processes was not necessary. Let the strength of induced absorption be $\beta$. Then the probability of induced absorption is

$$\beta \frac{N_\nu d\nu}{A_\nu + N_\nu d\nu} \tag{38}$$

Let the coefficient of spontaneous emission be $\alpha$. Then for equilibrium one gets

$$n_r \beta \frac{N_\nu d\nu}{A_\nu + N_\nu d\nu} = \alpha n_s \tag{39}$$

which agrees with (34) provided $g_r \beta = g_s \alpha$. A substitution of the value of $n_r/n_s$ then gives Planck's law. Bose then treated the cases considered by Pauli [1923] and Einstein and Ehrenfest [1923] and showed that the fundamental equilibrium condition (32) followed from the probability law (37) without requiring any further assumptions.

In a note added to the paper Einstein gave two reasons why he thought Bose's hypothesis about the probability of elementary radiative processes was not "applicable." In his letter of 3 November 1924 Einstein summarized the two reasons. "Your principle is not compatible with the following two conditions:

1) The absorption coefficient is independent of the radiation density.

2) The behaviour of a resonator in a radiation field should follow from the statistical laws as a limiting case." [Einstein to Bose, 3 November 1924]

Let us see to what extent these criticisms were fair. As far as the first point is concerned Einstein was referring to Beer's law. Bose's probability law, however, predicted a dependence of the absorption coefficient on the radiation density, decreasing with it. "If the behaviour had been such," wrote Einstein in the note, "then it would certainly have already been discovered in the case of infrared radiation from hot light sources." This is hard to believe. The departure from classical behaviour predicted by Bose's principle should occur only when $N_\nu/A_\nu << 1$, i.e. for very low intensity radiation. It is now well known that the departures from classical behaviour predicted by quantum optics are not seen even when the intensity of light is extremely low, unless the source happens to be of a special kind (like a single-photon source). The reason is that light from all classical sources is in a coherent state in which the number fluctuation is extremely large (Poissonian or higher). Therefore only light from "single photon states" (Fock states) and "squeezed states" and other such "sub-Poissonian situations" show departures from classical behaviour. [Aspect, 1987; London, 1983] Such light sources were certainly unknown in 1924.

The second point concerns the "correspondence principle" (not necessarily Bohr's which in this case implies that a radiation field can transfer positive as well as negative energy to a resonator, depending on the phase. "The probabilities of both these transitions must depend on the density of radiation, that is, on $N_\nu$ as opposed to Bose's hypothesis," wrote Einstein. The discovery of masers and lasers after the Second World War eventually vindicated Einstein's contention. It was nevertheless eventually an empirical rather than a sound theoretical refutation. Although the "correspondence principle" was used as a heuristic tool with great success by Einstein, Planck and later the Copenhagen school prior to the discovery of quantum mechanics around 1925/1926, it is no longer regarded as a fundamental principle. In fact, quantum mechanics itself is known not to satisfy this principle. [Liboff, 1980; Cabrera and Kiwi, 1987; Bohm and Hiley, 1985] For example, the nodes of the eigen-solutions inside a box do not disappear for arbitrarily high quantum numbers so that a classical particle cannot pass through such points. Also the cross-section for hard sphere scattering has no classical limit. Einstein was aware of this and used it against quantum mechanics as well.

## 15 Bose's Third (Unpublished) Paper (1925)

Although Bose was unhappy, particularly with the first point regarding Beer's law, he accepted the second point, modified his stance and wrote a third paper which he sent to Einstein from Paris on 27 January 1925. He sent him a letter under separate cover in which he wrote:

> It seems that the hypothesis of negative Einstrahlung stands, which, as you have yourself expressed, reflects the classical behaviour of a resonator in a fluctuating field. But the additional hypothesis of a spontaneous change, independent of the state of the field, seems to me not necessary...
> I am rather anxious to know your opinion about it. I have shown it to Professor Langevin here and he seems to think it interesting and worth publishing.

It was never published. Neither did Bose keep a copy of the paper, nor is there one in the Einstein Archives as far as one can ascertain. So, the precise contents and details will remain unknown. Bose arrived in Berlin in October 1925 and discussed his new ideas with Einstein. Einstein was so convinced of the necessity and reality of the two independent processes of emission of light from atoms (spontaneous and induced) that he could not agree with Bose that one of them, the former, was unnecessary to assume in deriving Planck's law.

To Bose, the process of emission of light appeared as a single physical process and the division into two independent processes appeared artificial. The assumption of one, he claimed, would automatically imply the other in thermal equilibrium, given the special statistical property of the radiation field. Initially Bose started with spontaneous emission as input but later changed his position. It was clear to Bose that Planck's law was a consequence of the special statistical property of the radiation field itself and was independent of the mechanisms of energy transfer with atoms and molecules.

This was indeed his capital contribution to physics and forms the substance of his first paper and the first part of his second paper. To Einstein, on the other hand, it appeared that Planck's law was a consequence of special mechanisms of energy transfer between the radiation field and molecules and that thermal equilibrium was impossible without both induced and spontaneous emission. This was indeed the lesson he derived from his 1916/1917 papers which he considered to contain "the fundamental outline of the future theoretical derivation."

Einstein was compelled to regard spontaneous emission as being independent of the state of the radiation field and an inherent property of atoms and molecules because in 1917 he was unaware of the special quantum statistical property of the radiation field. It is strange that even after championing Bose's method he failed to see Bose's point about spontaneous emission. To Bose it was clear that it depended on the "environment" in which the atom or molecule was placed. He wrote in his second paper [1924]:

> From classical theory one would expect that no stationary states are possible and that an interaction or coupling (Bohr) between ether and the excited atom and the radiation connected with it is always present. On the other hand, in order to explain spectral emission we must assume it to be possible that no interaction (emission) occurs. Because of the existence of stationary states we are therefore led to assign to every stationary state a probability coefficient or a mean life-time.

Recall the expression (37) for the fundamental probability $P$. It can be written as $\bar{n}/(\bar{n}+1)$ where $\bar{n} = N - \nu/A_\nu$ is the average number of quanta per cell. For $\bar{n} >> 1, P$ tends to unity, the classical limit in which stationary states are not possible. It follows from this that the average life-time of a stationary state (and therefore the probability of spontaneous emission) depends on the average radiation density $\bar{n}$. Bose used to recount to his students Einstein's objection to this. "Suppose there were only one hydrogen atom in the universe in an excited state," Einstein once remarked to Bose, "don't you think it would radiate spontaneously and come down to its ground state?" Bose did not disagree, but nevertheless, felt that such a gedanken situation did not capture the situation in a Hohlraum or cavity.

It was Bose's intuition that eventually turned out to be right in the end, not Einstein's. Firstly, in Dirac's method of "second quantization" (Dirac, 1927) which is widely used in quantum field theory today, one starts with a classical electromagnetic field in which the equality of the probabilities of induced emission and absorption (determined by the Fourier coefficients in the plane wave expansion) is built in, and obtains the spontaneous emission term as a consequence of the fundamental commutation rules. Although this is not exactly what Bose had proposed, it fits in with his basic contention that spontaneous emission is a consequence of the quantal property of the radiation field itself and need not be introduced as an independent hypothesis concerning the radiating atoms. Einstein's initial reaction to Dirac's contribution was also decidedly negative. [Pais, 1982, 441]

Secondly, in quantum electrodynamics spontaneous emission is not a property of an isolated atom but of an atom-vacuum system and can be significantly inhibited or enhanced by placing the atoms in a suitable environment. In quantum electrodynamics the vacuum is not empty and devoid of everything–it is full of virtual particles appearing and disappearing, interacting with one another; it provides a non-trivial physical "environment" to an atom.

A whole new branch of quantum optics called "cavity quantum electrodynamics" has developed since about 1987 utilizing dramatic changes in spontaneous emission rates to construct new kinds of microscopic masers that operate with a single atom and a few photons or with photons emitted in pairs in a two-photon transition. This is what two of the pioneers in the field have to say:

> Ever since Einstein demonstrated that spontaneous emission must occur if matter and radiation are to achieve thermal equilibrium, physicists have generally believed that excited atoms inevitably radiate. Spontaneous emission is so fundamental that it is usually regarded as an inherent property of matter. This view, however, overlooks the fact that spontaneous emission is not a property of an isolated atom but of an atom-vacuum system. The most distinctive feature of such emission, irreversibility, comes about because an infinity of vacuum states is available to the radiated photon. If these states are modified–for instance, by placing the excited atom between mirrors or in a cavity–spontaneous emission can be greatly

inhibited or enhanced.

> Recently developed atomic and optical techniques have made it possible to control and manipulate spontaneous emission. [Haroche and Kleppner, 1989]

In his third unpublished paper Bose had gone further. He wrote in his letter of 27 January 1925 to Einstein:

> I have tried to look at the radiation field from a new standpoint and have sought to separate the propagation of Quantum of energy from the propagation of electro-magnetic influence. I seem to feel vaguely that some such separation is necessary if Quantum theory is to be brought in line with Generalized Relativity theory.
>
> The views about the radiation field, which I have ventured to put forward, seem to be very much like what Bohr has recently expressed in May Phil Mag 1924. But it is only a guess, as I cannot say honestly to have exactly understood all he means to say about virtual fields and virtual oscillators.

This letter shows that Bose was one of the first persons to have formulated the idea of an "empty wave" (an electromagnetic wave propagating in space and time but carrying no energy-momentum). He was certainly the first person to have intuitively seen its relationship with the principle of General Relativity, although the connection is not clear. Einstein already had the idea of an empty wave in his light-quantum hypothesis according to which "a ray of light expands starting from a point, the energy does not distribute on ever increasing volumes, but remains constituted of a finite number of energy quanta localized in space and moving without subdividing, and unable to be absorbed or emitted partially." [Einstein, 1905]

If the localized quanta carry all the energy and momentum, what happens to the electromagnetic waves which can produce interference? There was no clear-cut answer. The problem was so acute that Einstein referred to these waves as Gespensterfelder (ghost waves) guiding the photons. [Bohr, 1945, 206] Louis de Broglie also had a similar idea in his concept of phase or pilot waves "guiding the propagation of the energy" and enabling a "synthesis of the waves and the quanta." [de Broglie, 1923, 549] Slater also arrived at a similar notion. As he recalled: "A number of scientists–W[illiam] F [rancis] G[ray] Swann, among others–had suggested that the purpose of the electric field was not to carry a continuously distributed density of energy, but to guide the photons in some manner. This was the point of view which appealed to me, and during my visit at the Cavendish Laboratory in the fall of 1923, I elaborated on it. [Slater, 1975, 9] Slater who believed in the existence of light-quanta as well as electromagnetic waves found the following way out of the difficulty: "I decided you could only have a statistical connection between them because you couldn't set up a vector to represent the photon in electromagnetic theory: that you would have the intensity of the wave governing the probability of finding the photon there. And I wanted to have the wave emitted during the stationary state so as to get it emitted over a long enough period so that it would have a suitable spectral distribution." [Slater, loc.cit., 30]

When Slater arrived in Copenhagen on 21 December 1923 to collaborate with Hendrik Kramers and Niels Bohr, he was prevailed upon by them to relinquish Einstein's light-quantum hypothesis but to retain his idea of electromagnetic waves emitted by oscillators during their stationary states and carrying no energy. They coined the name virtual oscillators for them.

The collaboration resulted in the famous Bohr-Kramers-Slater proposal [Bohr, Kramers, Slater, 1924] which Bose referred to in his letter to Einstein. In the BKS proposal the conservation of energy (and momentum) was abandoned for individual transition processes in atoms but was retained as a statistical concept. Most interestingly, they wrote: "the transitions which in (the Einstein theory of 1917] are designated as spontaneous are, in our view, induced by the virtual field." Bose must have noted this with satisfaction and this must have been one reason why he regarded the BKS proposal to be "very much like" his own.

The theory was however quickly falsified by experiments carried out by Bothe and Geiger [Bothe and Geiger, 1925] and Compton and Simon [Compton and Simon, 1925a, b] which established the conservation of energy and momentum in individual processes. Nevertheless, their ideas were reformulated and generalized by Born [1926] in his well known statistical interpretation of the wave function in quantum mechanics, an idea now universally accepted. This is what Heisenberg wrote about its historical antecedent:

> "The probability wave of Bohr, Kramers and Slater... was a quantitative version of the old concept "potential" in Aristotelian philosophy. It introduced something standing in the middle between possibility and reality.
> Later, when the mathematical framework of quantum theory was fixed, Born took up this idea of the probability wave and gave a clear definition of the mathematical quantity in the formalism." [Heisenberg, 1958, 29]

On 16 July 1946, twenty-one years after he received Bose's letter of 27 January 1925 from Paris, Einstein wrote to Schrodinger about the latest status of the unified field theory (of electromagnetism and gravitation based on the General Principle of Relativity):

> "It is indeed correct that the energetics occurs quite differently here compared to Maxwell-Poynting. I consider, however, that in this respect the Maxwell theory is really false on account of quantum actualities. In light the energy exists in something like quasi-singularities. The wave field as such should therefore not be the site of energy. One must simply require this and be happy that in the new theory the transverse wave field is indeed present but as such transports no energy." [Moore, 1989, 427]

There is no mention of Bose, his letter or his third paper which Einstein must have rejected outright, although the facsimile of Bose's letter supplied by The Jewish National and University library, Jerusalem, shows that he had marked the phrase "have sought to separate the propagation of Quantum of energy from the propagation of electro-magnetic influence" within parentheses.

Louis de Broglie's original interpretation of Schrodinger's equation [de Broglie, 1927] was given a firmer basis by Bohm [1952]. In the de Broglie-Bohm "casual interpretation" particles are thought to exist independent of observation, embedded in a wave field $\psi$ which satisfies the Schrodinger equation, with a density (over the statistical ensemble) of value $|\psi|^2$. One can show that this approach can explain the interference pattern observed in double-slit experiments without abandoning the notion of well-defined particle trajectories. [Philippidis, Dwedney and Hiley, 1979] Experiments for testing certain consequences of "empty waves" have been proposed [Selleri, 1990, ch.4] and some have already been carried out [Zou, Grayson, Wang and Mandel, 1992]. However, such experiments have not yet ruled out all versions of "empty wave" theories [Vigier and Holland, 1991].

Bose remained ever grateful to Einstein (whom he always addressed as "Sir" or "Master") for his encouragement and the interest he had taken in his first paper. Nevertheless, deep within himself, he felt a resentment that rankled all through his life. Three of his most perceptive and original ideas were scotched by Einstein:

1) The idea of the photon spin which is now established beyond doubt. (That Bose proposed the idea in his first paper in 1924 cannot as yet be established beyond every shred of doubt, although circumstantial evidences are strong.)

2) The idea that spontaneous emission is not an inherent property of an isolated atom, totally independent of the radiation field, also confirmed by experiments since 1987.

3) The idea of the propagation of electromagnetic influence carrying no quanta of energy, an area of current theoretical and experimental activity.

Bose could not publish a single paper during his two years in Europe at a time when physics was going through a turmoil and one of its most exciting periods. He returned to Dhaka a disheartened man and did not publish anything in theoretical physics for the next twelve years!

The author is grateful to N Mukunda, V Singh and J C Pati for a critical reading of the manuscript and suggesting improvements.

## References

A. Aspect, *Hyperfine Interactions* **37** (1987) 3-18.

David Bohm, *Phys. Rev.* **85** (1952) 166-179.

D. Bohm and B.J. Hiley, *Phys. Rev. Lett.* **55** (1985) 2511.

Niels Bohr, *Phil Mag.* (6) **26** (1913a) 1-25; *Phil Mag.* (6) **26** (1913b) 476-502; *Phil Mag.* (6) **26** (1913c) 857-875; *"Discussion with Einstein,"* in Albert Einstein: *Philosopher Scientist*, ed P.A. Schilpp, The Library of Living Philosophers, vol. VIII, Open Court, La Salle, Illinois, London: Cambridge University Press, Third Edition.

S.N. Bose, *Z. Phys.* **26** (1924a) 178-181; *Z. Phys.* **27** (1924b) 384-393.

Louis de Broglie, *Recherche sur la theorie des quanta.* Thesis presented to the Faculty of Science of the University of Paris for the degree of Doctor of Science. 1927 *J/Phys.* (Paris) **6** (1924) 225.

G.G. Cabrera and M. Kiwi, *Phys. Rev.* **A 36** (1984) 2995.

Ronald W. Clark, *Einstein: The Life and Times*, New York: World Publishing Company; paperback edition, New York: Avon Books 1972 (1971).

C.J. Davisson and L.H. Germer, *Phys. Rev.* (2) **30** (1927) 705-740.

P. Debye, *Ann. d. Phys.* **33** (1910) 1427-1434; *Phys. Zs.* **24** (1923) 161-166.

P.A.M. Dirac, *Proc. Roy. Soc.* (London) **A 112** (1926) 661-677; *Proc. Roy. Soc.* (London) **A 114** (1927) 243-265.

Paul Ehrenfest, *Sitz. ber. Akad. Wiss* (Wien) **114** (1905) 1301-1314.

P. Ehrenfest and H. Kamerlingh Onnes, *Versl. Lon. Akad. Wetensch.* (Amsterdam) **23** (1914) 789-792; English translation. *Proc. Lon. Akad. Wetensch.*

(Amsterdam) **17** (1914) 870-873; reprinted in Collected Papers, 353-356.

Albert Einstein, *Ann. d. Phys.* (4) **17** (1905) 132-148; English translation in D. ter Haar; *The Old Quantum Theory*, Oxford, London, Edinburgh, New york, Toronto, Sydney, Paris, Braumschweig: Pergamon Press 1967 (1905) 91-107; *Ann. d. Phys.* (4) **20** (1906) 199-207; *Verh. d. Deutsch. Phys. Ges* (2) **18** (1916a) 318-323; *Mitteilungen der Physikalischen Gesellschaft Zurich* **16** (1916b) 47-62; *Phys. Zs.* **18** (1917) 121-128; *Sitz. ber. Preuss. Akad. Wiss.* (Berlin) (1924a) 261-267 (presented at the meeting of 10 July 1924); *Abhandlung*, *Sitz. ber. Preuss. Akad. Wiss.* (Berlin) (1925a) 3-14 (presented at the meeting of 8 January 1925); *Sitz. ber. Preuss. Akad. Wiss.* (Berlin) (1925b) 18-25 (presented at the meeting of 29January 1925).

A. Einstein and P. Ehrenfest, *Z. Phys.* **19** (1923) 301-306.

Enrico Fermi, *Rend. R. Accad. Lincei* **3** (1926) 145-149 (presented at the meeting of 7 February 1926; *Z. Phys.* **36** (1926) 902-912.

W. Friedrich, P. Knipping and Max von Laue, *Sitz. ber. Bayer. Akad. Wiss* (Munchen) (1912) 303-322.

S. Haroche and D. Kleppner, *"Cavity Quantum Electrodynamics"*, *Physics Today* **42 1** (1989) 24-30.

Werner Heisenberg, *Z. Phys.* **33** (1925) 879-893; English translation in *Sources of Quantum Mechanics* (Van der Waerden, 1967), 261-276; *Physics and Philosophy, The Revolution in Modern Science*, Harmondsworth: Penguin Books, 1990 (1958).

Max Jammer, *Conceptual Development of Quantum Mechanics*, New York, St. Louis, San Francisco, Toronto, London, Sydney: McGraw Hill Book Co. (1966).

P. Jordan and E.P. Wigner, *Z. Phys.* **47** (1928) 631.

Klotz, *Jour. of the R.A.S. of Canada* **12** (1918) 357.

P. Langevin and Maurice de Broglie, *La Theorie du Rayonnement et les Quanta* - Rapports et Discussions de la Reunion tenue a Bruxelles, du 30 Octobre au 3 Novembre 1911, Paris: Gauthier-Villars (1912).

R.L. Liboff, *Phys. Today* **37** (1987) 50.

Fritz London, *Nature* **141** (1938) 643-644; *Phys. Rev.* **54** (1938) 947-954.

R. London, *The Quantum Theory of Light*, 2nd ed. Clarendon, Oxford (1983).

J. Mehra and H. Rechenberg, *The Historical Development of Quantum Theory*, **1** (1982) 1-2, New York, Heidelberg, Berlin: Springer Verlag.

R.A. Millikan, *Phys. Rev.* (2) **7** (1916) 355-388.

Walter Moore, *Schrodinger: Life and Thoughts*, Cambridge, New York, Port Chester, Melbourne, Sydney: Cambridge University Press (1989).

Ladislas Natanson, *Phys. Zs.* **12** (1911) 659-666.

S.B. Nicholson, *Monthly Notices of the Royal Astron. Soc.* **74** (1914) 425.

Abraham Pais, *"Subtle is the Lord..." The science and life of Albert Einstein*, Oxford, New York, Toronto, Melbourne: Oxford University Press (1982).

Wolfgang Pauli, *Z. Phys.* **18** (1923) 272-286; *Z. Phys.* **31** (1925) 765-783; *Z. Phys.* **41** (1927) 81-102.

C. Philippidis, C. Dewdney and B.J. Hiley, *Nuovo Cimento* **52 B** (1979) 15.

Max Planck, *Verh. d. Deutsch. Phys. Ges.* (2) (1900) 237-245; *Zur Theorie der Warmestrahlung.*, *Ann. d. Phys.* (4) **31** (1910) 758-768; *"La loi du rayonnement noir et l'hypothese des quantites elementaires d'action"* in Langevin and de Broglie (1912) 93-114; *1925 Sitz. ber. Preuss. Akad. Wiss.* (Berlin) (1911) 49-57.

C.V. Raman and S. Bhagavantam, *Ind. J. Phys.* **6** (1931) 353-366.

M.N. Saha, *Astrophys. Jour.* **50** (1919) 220.

Erwin Schrodinger, *Ann. d. Phys.* (4) **79** (1926a) 361-376; *Ann. d. Phys.* (4) **79** (1926b) 489-527; *Ann. d. Phys.* (4) **80** (1926c) 437-490; *Ann. d. Phys.* (4) **79** (1926d) 734-756; *Statistical Mechanics*, *A course of Seminar Lectures*, London: Cambridge University Press, Second Edition (1952).

F. Selleri, *Quantum Paradoxes and Physical Reality*, Dordrecht, Boston, London: Kluwer Academic Publishers (1990).

A. Sommerfeld, Attidel Congresso Internazionale dei Fisici. 11-20 Septembre, Bologna: Nicola Zanichelli (1928) **2** (1927) 449-473; *Z. Phys.* **47** (1928) 43-60; *Ber. dt. chem. Ges.* **61** (1928) 1171-1180.

G.P. Thompson, *Proc. Roy. Soc. A* **117** (1928) 600.

J.P. Vigier and P. Holland, *Phys. Rev. Lett.* **67** (1991) 402.

X.Y. Zou, T. Grayson, L.J. Wang and L. Mandel, *Phys. Rev. Lett.* **68** (1992) 3667-3669.

# Index

## P



## Q



## R



## S



## T



## U



## V



## W